"David Houle has seen the future. rest of us on his ever-accelerating bike to show us how this coming decade will be more fraught with change than any two or three previous decades combined. Embrace the opportunities ahead! A must read."

— Robert J. Sawyer, Hugo and Nebula Award-winning science-fiction writer

"In his series of books on the 2020's David Houle aims to raise our sense of urgency about the issues we face. This book captures the internal conflict we feel when we compare this urgency and the relative lack of actions being taken. Houle skillfully urges the reader not to turn away but to confront our challenges head on. A succinct and compelling read."

— Glen Hiemstra, Futurist and Founder of futurist.com

"Futurist David Houle's book "The 2020s: A Decade of Cognitive Dissonance" encourages us to be open to a future decade that will challenge our attitudes and belief systems. David Houle's insights of the future will take you from Cognitive Dissonance to Cognitive Consonance — showing a path to harmony of what the future could bring to help you adapt and build your future in the face of the Abnormal 20s that could become the Soaring 20s if put into practice."

— Doug Hohulin, 5G/6G, AV, XR Technologist & Futurist - Nokia Solution Architect

"As a college president I have had to manage massive amounts of change and expect that trend to continue in the years ahead. I really like this book as it focuses on how to position ourselves to successfully navigate change. I recommend it to anyone bracing for the disruptive decade of the 2020s."

— Carol F. Probstfeld, Ed.D. President, State College of Florida, Manatee / Sarasota

"Get Ready! We are going to living through a decade of unprecedented change and disruption. If you want a preview of what's ahead — this is an essential guide- book for the coming decade. It provides wisdom, vision and clarity to give you the conceptual context needed for the tsunami of change directly ahead."

— Yanik Silver, author Evolved Enterprise and founder Maverick1000

"As we face unprecedented times at the start of this decade and the acceleration of technology continues its blistering pace, I worry for humanity that we are unprepared and lacking true leadership to meet the challenges before us. Step 1 is to understand and get context to where we truly stand, what we are dealing with and what the trend lines show and what that means for us individually and collectively. David Houle helps us gain this understanding better than anyone, through thought provoking, insightful and research-backed analysis. This second installment of "The 2020s" series is mind blowing yet very matter of fact."

— Arman Rousta, Founder of b.labs Ventures

The 2020s:
A Decade of Cognitive Dissonance

Book 2

David Houle

ISBN: 978-1-7339029-3-9

Dedication

To those I love, may love grow throughout the 2020s.

To those I have called good friends through the decades, may you live to see a preferred future and pass it to those who follow.

To all those visionaries, futurists, science fiction writers and forward thinkers who realize that our immediate collective future can be a transcendent one if we let go of the limitations from the past.

Table of Contents

Acknowledgements

First I want to thank Bob Leonard, the editor of this book. Bob is a fantastic editor and a patient one. He was my co-author on "Moving to a Finite Earth Economy – Crew Manual". We have worked together creating the books and columns that we feel are essential for those concerned with our climate crisis, the need to reinvent capitalism and the transformative possibility of a new emerging consciousness. Know you are valued.

Dave Abrahamsen and his design firm What2Design has been my creative support for the last 15 years. He has designed, created and maintained all of my websites, designed the covers of my books – including this one – and makes me look good to the eyes of the world. I always wait with excitement to see his designs and visual ideas.

I want to thank my colleagues at The Sarasota Institute – a 21st Century Think Tank for their support and intellectual stimulation… including the entire Advisory Board for their advice and support; and Carol Probstfeld and Bill Johnston who agreed to be Members of the Board of the new non-profit iteration of the Institute.

I want to thank all the creative minds at the Ringling College of Art + Design, where I am Futurist in Residence, particularly Larry Thompson, Doug Chismar and Tim Rumage. Larry is a true force of nature and the reason that Ringling College is now one of the top art and design colleges in the world. Doug showed me the limits of my knowledge with kind suggestions and recommendations. Tim taught me more about the environment and our climate catastrophe than anyone else, and is the top planetary ethicist I know on Spaceship Earth.

I want to thank the inspirational collaboration of my two fellow initiators of the #forkintheroadproject – Gerd Leonhard and Glen Hiemstra. Futurists all, we share a vision of the fork in the road that humanity is facing in the 2020s, and agree it is both urgent and an opportunity to create something better. There's still time for humanity to co-create the road to a preferred future rather than mindlessly following the path we are currently on. See Chapter 9.

Then there are all the intellectual, visionary, spiritual and inspirational minds who have shaped who I am and thus are in some way, through me, a part of this book:

Alvin Toffler, Marshall McLuhan, R. Buckminster Fuller, Teilhard de Jardin, Bob Dylan, the Rolling Stones, Led Zeppelin, Isaac Asimov, Robert Heinlein, Frank Herbert, Bruce Lee, Bobby Kennedy, Ram Dass, the Dalai Lama, Jack Kerouac, Henry Miller, Ernest Hemingway, John D. McDonald, Miles Davis and innumerable blues and jazz artists whose names would double the length of this paragraph.

I am sure I have forgotten many whose names should be included here. I apologize. Sometimes the memory may falter, but the vision of the future remains clear.

Quotes

"We should try to be the parents of our future, rather than the offspring of our past." — Miguel de Unamuno (1864-1936), Spanish essayist, novelist, poet, playwright and philosopher

"The illiterate of the 21ˢᵗ century will not be those that cannot read or write. It will be those that cannot learn unlearn and relearn." — Alvin Toffler (1928- 2016) American Futurist

"Those not busy being born are busy dyin…" — Bob Dylan, singer-songwriter, Nobel Prize winner for Literature

"You never change things by fighting the existing reality. To change something, build a new model that makes the existing model obsolete." — R. Buckminster Fuller (1895-1983) architect, designer, futurist, inventor, scientist

Prologue

The final draft of this book was written in the second spring of COVID-19. This means that the concept of Cognitive Dissonance has been experienced by literally billions of people in the last 12 months. Our collective reality, and our individual realities, have been abruptly and radically altered. Except for centenarians, no one has experienced anything like this pandemic reality before now.

In the 10 months since Book One of this 2020s series was published, my opportunity to help people think about the future, and advise them on ways to embrace significant change, has grown. Prior to the pandemic, it was often difficult to loosen the grasp of audiences on what they believed to be their present reality… to be open to see what is ahead and imagine all the changes and disruptions rushing at us. I would have to ask them to consciously suspend their concepts of reality. This was due to the inclination of people to think that their reality is THE reality.

Since the summer of 2020, my role as a futurist became easier in this regard. Now, when I speak to an audience, or advise a business, they are in the midst of a changed reality they did not see coming. Having had their "reality rugs" pulled out from under them, they now have a visceral acceptance that reality is mutable. "What is going to happen next?" is a question that comes from wonder and a place of not knowing… a terrific place for an open mind.

In the last year, people have had their minds opened to the reality of change. This is a auspicious starting point for the 2020s!

During this past year I have written many columns on my blog, on Medium, and in my bi-weekly newspaper column "The Futurist". One of them has been expanded into the second chapter of this book (which concerns how COVID-19 prepares us for the 2020s). Several others touched on the absurdity of the question, "When will things get back to normal?" First there is no normal, never was and never has been. Second, what was is not something we can go back to (except via recollection).

A metaphor we are hearing often as I write this (during a time of widespread vaccinations) is that we might be seeing "the light at the end of the tunnel". Think of that metaphor in this way: the daylight we are approaching is not the daylight we left when we entered the tunnel. It is a different daylight, a different time, a different reality. Reality is always in a state of flux. So when we exit the tunnel, reality will have changed.

What COVID-19 truly is, from the perspective of the reader of this book, is both preparation for a most disruptive decade, and an accelerant for change. During this past year, in speeches and columns, I have argued that during this past year of the virus, the trends creating our future have accelerated. "Five years of change collapsed into one year."

In addition to accelerating change, it has conditioned us to adapt to the reality of disruption. That reality is much easier to grasp now than in, say, January of 2020.

This is why I wanted to make this book the second in a series about this most momentous decade. Living within the state of cognitive dissonance is the new reality for the 2020s and beyond.

"Woke" is a new word in our political and social lexicon since the Black Lives Matter global demonstrations of summer 2020. "Woke" now has a pejorative meaning: a hard core enlightened view of social issues, often felt by those who are not victims, speaking on behalf of the victims.

Please realize that a deeper meaning of the word "woke" is that we now have all – to different degrees – been shaken awake by these profound disruptions. This book will help you live with a deep sense of cognitive dissonance as we hurtle toward a collective future that has within it the opportunity for creating our "good" future... rather than the "bad" one our current path all but guarantees.

We are at a collective fork in the road (Chapter 9). We need to move to thinking of humanity as WE not Us vs. Them (Chapter 10). May this book help you to accept cognitive dissonance and learn to flow with it .

Introduction

The 2020s is the most disruptive and transformative decade in history. There is no major part of society, culture, business, economics, politics and civilization overall that will not be disrupted and changed. In fact, it is these next ten years that will set the direction of civilization for the rest of the century.

This is the second book of a series that will look at the major trends, forces and disruptions of this decade. This book provides an understanding of what will be an almost constant state of change. There is no going back.

On many fundamental levels what we are entering is a completion of the Shift Age, the transition from the historical reality humanity has experienced up to 2000 to the new, altered reality we will be largely living in by the 2030s.

Cognitive Dissonance is a phrase many do not know the definition of, yet most of humanity is now in this state. This book is an attempt to educate about this unmoored feeling of disorientation and to help prepare us all for learning how to live in it... and create our future within it. Part of successfully adapting to massive change is to have some forethought about it, and what it might mean.

For those of you who know my work, read my books, blogs and columns or have heard me speak (in person or on video), thank you for returning. For those of you who are reading a book of mine for the first time, here is a brief introduction. After living a life that, in many cases, was "ahead of the curve" (see my bio at the end of this book), I became a full-time futurist in 2005. As a young man I read a lot of science fiction and books by futurists. The three most influential futurists of the 20th century – R. Buckminster Fuller, Alvin Toffler and Marshall McLuhan – shaped my thinking and continue to do so. In the most honorable way, I stand on the shoulders of these three great visionaries to look further into the 21st century than they did. To a great degree this is because their clarity of thought and the sheer size of their intellects is still of great value in navigating our future.

As we have learned in the last 50 years, science fiction often becomes science fact. This will accelerate in the 2020s. Hand-held devices, body scanners, brain implants, space travel, driverless vehicles, a dramatically changed definition of the workplace, entirely new business models, life expectancy extension, genetic enhancement, and a new growing collective consciousness... are all things that will make their continued impact in the 2020s. Fiction becomes fact. Imagination becomes manifest in the real world.

We stand at the apotheosis of human evolution. We are smarter, bigger, more technologically integrated, wealthier and better educated than at any time in history. In the last two centuries, the Industrial Age and then the Information Age have delivered unprecedented progress in the quality of our lives.

Standing at this historical peak of human evolution, we have a true chance to take our next evolutionary leap. We can successfully navigate the dynamics rushing toward us to realize our ultimate next step: a merging of humanity and technology that can springboard humanity to a new level of abundance, higher consciousness and a global integration never experienced before.

However, the 2020s might also be a decade of reckoning… across the board we will be facing "fork in the road" decisions as a species that will absolutely determine the path of human civilization for the rest of the century.

The coming decade will shape the future in ways almost unimaginable.

Let's take a look!

Chapter One – What is Cognitive Dissonance?

The first step in understanding why the 2020s is a decade of cognitive dissonance is to understand the dictionary and encyclopedia definitions of the term.

Here is Webster:

- psychological conflict resulting from incongruous beliefs and attitudes held simultaneously.

Dictionary.com:

- anxiety that results from simultaneously holding contradictory or otherwise incompatible attitudes, beliefs, or the like, as when one likes a person but disapproves strongly of one of his or her habits.

Encyclopedias provide deeper definitions.

Wikipedia:

- In A Theory of Cognitive Dissonance (1957), Leon Festinger proposed that human beings strive for internal psychological consistency to function mentally in the real world. A person who experiences internal inconsistency tends to become psychologically uncomfortable

and is motivated to reduce the cognitive dissonance. They tend to make changes to justify the stressful behavior, either by adding new parts to the cognition causing the psychological dissonance or by avoiding circumstances and contradictory information likely to increase it.

Here are some quotes about cognitive dissonance to provide a better understanding of the concept:

"Sometimes people hold a core belief that is very strong. When they are presented with evidence that works against that belief, the new evidence cannot be accepted. It would create a feeling that is extremely uncomfortable, called cognitive dissonance. And because it is so important to protect the core belief, they will rationalize, ignore and even deny anything that doesn't fit in with the core belief." – Franz Fanon

"When dissonance is present, in addition to trying to reduce it, the person will actively avoid situations and information which would likely increase the dissonance." – Leon Festinger

"Wisdom is tolerance of cognitive dissonance." – Robert Thurman

Our experience of dissonance in the 2020s will require continued management of "reality" fading into the past simultaneously with new realities showing up with accelerating frequency. A psychological opening up to change that forces willingness to accept and then embrace change.

These definitions from dictionaries and encyclopedias need to be expanded and recast to the bigger dynamics at play. In traveling the world giving keynote speeches and presentations, I have asked audiences to suspend what they think of as their current reality. I have asked thousands of people to, at least temporarily, suspend what they think reality is, so they can be open to what I suggest is the future. While people understand that change is real, they often have a hard time adapting to it… they assume that it won't affect them if they don't want to change.

It will affect them, and eventually they will be forced to change.

Reality is never fixed, but most people think that it is. Some of this is habit. Some of this is assumptive belief that today will be much like yesterday and that tomorrow will be much like today. "I will get up, get ready for work, commute to work, spend a day working and then come back home. Repeat. The sun always rises and sets. I will wake up in the same body I went to sleep in."

We then take this familiar repetition and project it into the future. We do this assuming that what lies ahead is just an extension of today's perceived reality… just as today's is an extension of yesterday's.

The metaphor here is rowing a boat. We move toward our destination, but we are facing where we came from, not where we are going. Every now and then we need to turn and look to see if we are on course. We are backing into the future.

In the 2020s, we must turn and face our future.

We can no longer think that past is prologue (except from the viewpoint of time sequence). The past of the last 20 years brings us here, but it in no way prepares us for our day-to-day reality of the next decade. Our present will be disorienting, and our future will be full of disruptions. Our past is not an accurate indicator of the future we are accelerating into.

The cognitive dissonance we will feel in the 2020s will be greater than for any other iteration of humanity. This is a result of the number and magnitude of the disruptions we will experience. Later in this book I will cite past eras that experienced a similar degree of change... but over 50 year timelines. The ten years of the 2020s will experience at least as much change (massive, unique and aggregate change) as any 50 year period in history. We will all have to learn to adapt and be open to a degree of change and disruption that no previous iteration of humanity has experienced.

The massive amount of change, the accelerating speed of change, and the unprecedented simultaneity of global change in the 2020s is a first time event. People will become disoriented as they struggle to adapt. There will be significant cognitive dissonance for most people as what their assumed and expected "reality" ceases to exist.

In Book 1 of this series on the 2020s[1] the Shift Age was discussed. I have written several books about this age that we now find ourselves in.[2]

The Shift Age — following the Agricultural, Industrial and Information ages — began in the 2000s and will continue until the 2030s. It is a time of transition from the reality humanity lived in up to the year 2000 to the new reality of the 2030s.

Literally.

Think about how you have experienced, embraced and adapted to technology in the last 15 years... and how much it changed your life. Here are just a few of the new technologies that we have learned to accept:

- cloud computing
- computer interface transitions from keyboards to touchscreens to voice
- eBooks
- electric vehicles
- global video calls from a hand-held device
- high-speed wireless
- laptops – ever more powerful, lighter and cheaper. Anyone can buy a computer for a few hundred dollars today that exceeds the computing capability of super computers of the 70s and 80s costing tens of millions of dollars.
- renewable energy (has increased from 1 to 11% of all electrical energy used globally)
- smart phones
- social media
- the growth and acceleration of Artificial / Technological Intelligence
- the migration of retail to the ability to purchase anything online

- video and audio streaming services.

Now, consider this: the current speed of technological change is the slowest it will be for the rest of your lifetime. It is only accelerating.

To some degree, accelerating technological disruption is an early warning system for the disruptions directly ahead in all facets of life during this decade.

2020 was obviously the year of COVID-19. We will take a deeper look into the virus in the next chapter as it is a worthy metaphor for the rest of the decade. Let's take a quick look now as to how the virus jump-started cognitive dissonance for us all.

The two realities of the Shift Age are the Physical Reality and the Screen Reality. We will review these closely in Chapter 4, but for now it is an appropriate lens to see why COVID-19 has provided us with our first huge cognitive-dissonant event.

We all entered 2020 with some excitement, curiosity, and for many, apprehension.

As we reached March, it became clear that we had entered the era of COVID. The vast majority of nations went into some form of lockdown and self-quarantine for several months. Our reality changed overnight. We were in our homes… doing all the things we usually do at home, but we couldn't venture out. Stores, restaurants and places of work were closed. Billions started working from home. Zoom, what is Zoom? How do I use it? How do I work from home? What am I supposed to do with my school age kids who are now learning from home? Do I wear a mask? Do I clean every surface all the time? What does it mean to shop, communicate, organize, teach, have medical appointments all online?

Our individual and collective realities changed radically in a matter of days. Remember all that? That was when our world plunged into a state of cognitive dissonance.

Remember those feelings. Remember that disorientation. Remember how you adapted. What was the most difficult part of it for you? That will be different for different people, and it is helpful to be aware of what you found most distressing. It will serve you well because COVID-19 is just the first of many disruption to come that you will need to manage on several levels.

Pre-COVID19, I traveled the world giving speeches. Whether it was a boardroom with an audience of 15, a corporate retreat for 30, or a conference with 5,000 attendees, the experience was largely the same. I drove to an airport filled with people, flew to another airport on a plane filled with people, then perhaps another airport, another plane, then a limo or Uber to a hotel where I usually attended a pre-conference reception, shaking dozens of hands, having numerous up close and personal discussions. This was followed by a keynote speech delivered to a room full of people, many of whom spoke with me after the speech or at the book signing table. Then I traveled back home… flying over 200,000 miles a year.

Then the virus drove us into lockdown.

My business model was gone. Gone for at least two years, assuming that a vaccine would be developed, approved, and distributed to 75% of the population by the end mid-2022. What do I do? Retire early? No. What I did was I reinvented my business model. My guiding light was this quote from Bruce Lee, the martial artist and movie star:

"Don't get set into one form, adapt it and build your own and let it grow, be like water. Empty your mind, be formless, shapeless, like water. You put water into a cup, it becomes the cup, you put water into a bottle, it becomes the bottle, you put it into a teapot, it becomes the teapot. Be water, my friend."

So, I did. I viewed the virus as my new container. I viewed COVID-19 as the new vessel or universe in which I had to operate. I flowed like water into this new container, and then started to adapt to it and create within it.

I created a green-screen virtual studio in my house to replace my travel intensive pre-COVID business model.

I have continued to speak to business groups, and a number of individual CEOs and business owners. They turned to me as a futurist to help them contextualize the new decade and the virus itself. Instead of flying to them, I Zoomed with them via the screen reality. My new business model has lower revenues, but also lower overhead expenses… and much less time spent traveling.

I advise my clients to get comfortable being in the state of cognitive dissonance. It helped that I could show them how I had adapted. Successful in doing this, I was able to provide my personal experience as an example of the flexibility, resilience and adaptability they must adopt to successfully weather the 2020s.

COVID-19 changed things quickly and dramatically. It also accelerated many of the societal and technological transitions that were going to happen anyway, but now faster and sooner.

Let's take a look at the virus as the first massive disruption of the decade.

[1] "The 2020s: The Most Disruptive Decade in History"

[2] "The Shift Age", "Entering the Shift Age", "Shift Ed" and "Brand Shift"

Chapter Two – Why COVID-19 is the Bike with Training Wheels for the 2020s

The COVID-19 pandemic may be the single greatest cognitive dissonance event in my lifetime. During the Spring of 2020, billions of people simultaneously started to live a new reality… one of self-quarantine.

We were in our familiar homes with our families, loved ones or close friends, so our immediate surroundings were the same, but the reality outside our dwellings had changed. Inside was safe, outside was not. The disconnection was profound. Given that this was a new virus, all its dynamics and effects could only be learned through time. Virus transmitted via surfaces. Clean all surfaces! Don't wear a mask! Virus transmitted through the air! Wear a mask! What are the metrics for hospitalizations and deaths? What is an acceptable positivity rate?

Endless questions and our health perceived to be at risk.

COVID-19 taught us that what we thought of as reality at the start of 2020, was no longer. We had to live in our continuing reality with a new one overlaid on top. How do we manage our lives, jobs, homes etc. in profoundly different ways?

Disruption. Disorientation. Confusion. Cognitive dissonance.

This was the common experience around the world. All of us were thrown into a life changing situation for which we had no prior experience. Learning a new reality while trying to manage the one we had (thought was THE reality), but no longer was.

For billions this was an ongoing, simultaneous management of cognitive dissonance.

Riding a Bike with Training Wheels

Clear image, right? Either a memory from your childhood or from parenthood. The first lesson is to learn how to pedal, brake and turn. Then learning the hand signals for traffic, and other safety rules. Then the training wheels come off and the requirement for balance is the next big lesson. Once learned, always remembered. That truth has become a metaphor for whenever someone says they haven't done something routine in a long time. One says, "I haven't done that in years!" The other responds, "It's just like riding a bike!" Everyone reading these words has experienced this dialogue.

What else do we know from riding a bike? That one cannot stay in the same place; she must move forward!

"Life is like riding a bicycle. To keep our balance you must keep moving." – Albert Einstein

The 2020s is the most disruptive decade in history. COVID-19 is one big global disruption. Over the past year, everybody has been disrupted to one degree or another. We all had to make fundamental changes in how we live. We all are still living differently than we did this time a year ago. Healthcare workers, hospitality/travel/retail and just about anyone in the field of education is in full on adjustment mode and largely making it up as they go. Initially, these groups of workers had to fight leaders making decisions based upon politics not science. They had to struggle forward with little or no funding or support. Politicians were not adapting to the new reality, but continued to respond as if we were still living in a pre-pandemic reality. Leaders in the 2020s must live within cognitive dissonance just like the rest of us.

Business owners, large and small, have either closed their doors for good, or are working like crazy to keep from doing so. We all know the issues and the challenges. What most don't understand is that we will never return to normal.

What will the new normal be? When will everything go back to how it was? It won't.

There is no normal. The only "normal" is abnormality. Be open to, and even embrace, abnormality. In a column published in June 2020, I wrote:

Please rise above the group think that the media spews at us. Please don't use the term "new normal". There will not be anything remotely normal for quite a while. Get comfortable with the abnormal. Abnormality is the "new normal". The only constant in the universe is change. The difference today, in this first year of the 2020s, is we must adapt to comprehensive and disruptive change. By the end of this decade, it will have creatively destroyed legacy thinking and collapsed our legacy reality.

The single largest psychological symptom of the pandemic is the belief that at some point "this" will be over. It will not.

The theater business, the restaurant business, the office leasing business, the concert business, the in-person conference business and all the other businesses that are based upon large groups of people coming together in a space will never be the same. Some of these businesses will continue to exist, but many fewer than before COVID-19. Those that remain will be different in the way they operate and how they serve customers.

False hope can be deadly. Please don't hang on to it.

Why is almost everything going to be different even when COVID-19 goes away?

The fact that America is the plague nation of the world puts us in a hole. The previous administration's corruption of the scientists at the CDC, the FDA, and HHS has resulted in widespread distrust of the government.

I have heard many variations of this statement: "Let all our political leaders take the vaccine first. If nothing bad happens to them it will be good for the country. If something bad does happen to them, it will be good for the country."

That reveals the level of distrust people have of those in power, who have not led but only sowed confusion. This means that, at a time when many are reticent to take a vaccine, to attain herd immunity (and for us to have a viable post-COVID19 economy), an effective vaccine must be administered to 75% of the population. This trust issue is another example of cognitive dissonance.

The lag time from the inception of a disruption to effective action on the part of global political leaders will continue. These leaders came to power during realities that are being disrupted. They have no prior experience to tap into. This will provide more cognitive dissonance as leaders will largely not be able to lead in the traditional definition of the word.

Second, over time people change. Particularly when they have experienced a major disruption. Research shows that new habits can be fully formed in several weeks. We have had months to recalibrate our lives. Streaming services have exploded during COVID-19, with monthly subscriptions costing less than the price of a single movie theater ticket. We may go back to attending 20th century physical theaters, but not all of us, and not with the same frequency.

We have learned to cook and eat at home much more frequently. We spend less on food. The sourdough bread phenomenon has been well-documented. We have become cooks, dieters and bread makers. Yes, we will go back out to restaurants, but less often, and more just for special occasions.

The largest disruption has been working from home. Millions who never did, do now. Many like it so much that they don't want to go back to the office and the work commute ever again. Companies see the cost savings they have realized and support their employees desire to continue working from home. Look for vacancies and darkness in downtown office buildings across the world.

COVID-19 has given us the opportunity to learn how to ride a bike with training wheels. We had never ridden one and now we have. However, we haven't yet learned how to retain our balance in the midst of on-going disruptions. Very few of us have established a state of balance with the pandemic. Many feel that things will be off-kilter "until we return to normal". That's a delusion.

In the 2020s there will most likely be another global pandemic. In addition, COVID-19 is showing us that it can adapt, morph and reconstruct itself quickly. This suggests that we will be dealing with various strains of the virus for years to come. The good news is that we have learned the rules of riding the virus bike: wearing masks, social distancing, washing hands and surfaces more frequently.

There will be deflation in some parts of the world and inflation in others. There will be a massive "reckoning" over debt and finance. If 100% of America's annual GDP went to pay down the federal deficit, we still would not retire all of it. Think about that! There will be unprecedented numbers of unemployed and under employed.

There will be major geopolitical upheavals, and continued wealth inequality. And the biggest disruptor of all, our climate crisis, is on our doorstep. Have you watched the news lately? In a recent book, I forecast a range of 50 to100 million climate crisis refugees in the next ten years, globally… and probably at least five million of those in the United States.

This Has Never Happened Before!

Something happened in the spring of 2020 that had never happened before. Billions of people did the same thing at the same time. In every country, to a greater or lesser degree, people self-quarantined for 45 to 90 days. Air travel dropped 90%, the price of gasoline dropped to its lowest levels in years as people stopped driving, and all forms of in-person entertainment and conferences shut down.

One of the most significant developments was that while humans were all self-quarantining during the spring, nature and the environment had a respite. For the first time in decades, people in Northern India could see Mount Everest as the level of pollution plummeted. The dirty canals of Venice became clear, and fish returned. All across the world, animals came out and freely traveled through human landscapes, reclaiming habitat that had been theirs.

One significant measurement of this was that, for the first time in 50 years, Earth Overshoot Day (EOD) moved out a month. This is the day after which humanity is taking more from the earth than the earth can replenish on an annual basis. Balance requires that EOD be on 12/31 of any year (meaning that the human species did not go into deficit with the planet's resources). In 2019, EOD was July 29th, and it was projected that the date for 2020 would be July 26th. After the global lockdown, when, for months, we stopped activities that put GHG emissions into the atmosphere, EOD moved forward to August 22nd – almost a full month!

This proved several things. First, that indeed we are the cause of pollution, environmental degradation and climate change. Second, in order to face our climate catastrophe [see Book 1 in this series and a future book on the topic], we may have to systematically institute lockdowns to temporarily slow rapidly escalating climate-related disasters. Third, the major critical issues of the 2020s will require collective human action... coordinated simultaneously.

COVID-19 gave us a chance to learn how to ride a bike with training wheels. We all now have to learn how to ride through disruption after disruption and keep our balance while doing it.

Welcome to the 2020s!

Another significant dynamic that extends the bike on training wheels metaphor was the global magnitude of the pandemic disruption. For the first time in human history, three to four billion people all did the same thing at the same time! That had never happened before. The numbers were staggeringly large. More people simultaneously self-quarantined than were alive during WWII. The percentage of humanity that self-quarantined was 50 to 60% of the total population. Never has that large a percentage ever done anything simultaneously.

This was the best preparation for the rest of the 2020s. One reality of this decade is that in facing multiple disruptions, humanity will have to act in a global, coordinated fashion for extended periods of time.

Three to five billion people were in some level of self-quarantine. We stayed in place. We only went out to shop for essentials. We bought online. We used new delivery and pick-up services. We learned Zoom. We signed up for streaming services. We adjusted to working at home. We grew to appreciate teachers and frontline workers in new ways.

As stated earlier, we realized that "reality", as we knew it, is no longer. This is key. As a futurist, I have always opened keynotes and presentations by asking the audiences to "please suspend what you think reality is, as you will not be able to be open to the future if you think that the current reality will persist."

Just compare what you thought reality was in January of 2020 to right now. Is it the same? Of course not.

The ONLY constant in the Universe is change. If there was no change, there would be no time. Reflect on that. Reverse it. If there was no time, there would be no change.

This is obvious to someone who researches and detects the direction of the next five to ten years. After this moment, the future is the only place we have to live. It is in this moment that the future can be glimpsed. To paraphrase Marshall McLuhan:

"Most people drive down the freeway of life looking in the rearview mirror."

There isn't one of you reading this who can't tell me about your past. Tell me about your future. What do you see?

In our collective past is this big thing we did together. Over the past year, most of us only went out when absolutely necessary and we wore masks when we did have to go out. Essential workers had to go out every day. Others went out maskless because they listened to leaders who are either ignorant or who have priorities above human well-being.

We completely changed how we lived. Billions changed how they lived for the collective common good. How remarkable!

What this means is that we can do it again. And we will need to take big, collective actions many more times during the 2020s.

Our climate crisis demands such actions. They are crucial in this decade if we want to preserve our civilization.

Utopia or Oblivion?

"Whether it is to be Utopia or Oblivion will be a touch-and-go relay race right up to the final moment… Humanity is in a final exam as to whether or not it qualifies for continuance in the Universe." – R. Buckminster Fuller

Fuller wrote this back in 1969, an incredibly productive year for him as he published two books: "Operating Manual for Spaceship Earth" and "Utopia or Oblivion: The Prospects for Humanity".

These two books shape the essential thinking for humanity in the Age of Climate Change and the disruption of the 2020s. These two books, along with others by Alvin Toffler and Marshall McLuhan, supplied the foundation of my thinking as a futurist. I have frequently shared from stages around the world that, "I stand with honor on the shoulders of the three greatest futurists of the last 75 years to better see into this new century."

The books written by these three deep thinkers are perhaps more relevant today than almost anything that is currently being published… at least from the macro point of view. Fuller looks at the big picture of the future of humanity. McLuhan provides us with the reality of living in an electronic village. Toffler predicts how culture, business and society will evolve.

All three are still the most accurate, big picture visionaries in print. Most of the rest of us live in the worlds they envisioned.

Utopia means that we move forward in an ever more inclusive effort to get billions of people to collectively commit to taking actions that halt the warming of our atmosphere: a rapid transition to clean energy, a worldwide effort to educate girls because (amongst many other positive outcomes) it reduces birth rates, an adoption of regenerative agriculture, and many others.

Last Spring we did it. We can do it again. If we do, then the path to Utopia becomes paved in promise and abundance... powered by our collective will.

Clean energy.
Educate Girls (+ reduces birthrate)
regen regenerative agriculture

Powered by Collective will

Chapter Three – The Collapse of Legacy Thinking

Legacy thinking is beliefs and world views from the past. The thinking that guided actions in the past and, so far, has mostly worked. The thinking that led to the creation of recent and current realities. Thought precedes action. The thinking of the past led to the reality of today.

All through history, legacy thinking has collapsed and been replaced by new beliefs as a result of creative destruction or external forces.

Whenever this has happened, it was at first not recognized as long-held beliefs and concepts kept people from seeing the change. Humans hold on to what has been reality. They don't see change coming, and only notice it when it can no longer be avoided. That is why there has always been some degree of cognitive dissonance. Held beliefs continue until a new reality can no longer be denied. As the great science fiction writer William Gibson stated:

"The future is here… it is just not evenly distributed."

Change normally first occurs in a single place, then several places and then spreads beyond these points of genesis. Through time people adapt and accept the change. The profound difference with the 2020s is that the speed of change is now environmental. We will not have time to move gradually into the new future. It will feel sudden.

The collapse of legacy thinking will trigger cognitive dissonance as what was perceived as reality, no longer is.

What was, what reality was and all the thinking that brought it into being, is dissolving in the present as we transition to what will be. Past moorings to reality still exist, but they are giving way to entirely new forms of thinking, beliefs and relationships. Fixed is becoming mutable. Change is constant and ongoing. Attachment to legacy views will create blind spots into seeing what is actually going on.

This collapse of legacy thinking is something I first explored in my 2013 book "Entering the Shift Age". The concept became one of the most referenced topics of the book. At the time of writing, I didn't expect that. Sometimes what resonates with readers is not fully anticipated by the writer.

To quote from the book:

"Legacy thinking is often why it is so hard to see change, especially now because we are living in the present through the filters of the past. Only when the change becomes personal do we understand the change. And then there is the conflict: either you have to accept that your legacy thinking is no longer valid, or you hold on for a more secure and comfortable view of the world. That is why people get so upset by change

— it threatens their point of view or worldview. Take a moment to review your own thoughts and how you use them to evaluate and see the world around you. Are you looking at what is going on in 2012 through the filter of concepts or thoughts that you accepted as valid years or even decades ago?"

Drop in "2020" for "2012" and this is even more true. Think of all the things going on that force you to struggle with your view of the world. The upset you feel due to the virus and the resulting social and political upheavals (e.g. the tribal disagreements over wearing masks) at its foundation is the feeling that things are "out of control". That's simply because they don't fit what you think reality is. What reality **was.**

Examine the thoughts you have around almost anything, and you will see that you are confining your view of the present to the keyhole of your legacy thinking.

- Do you still hold onto beliefs, taught to you in childhood, that you are superior because of your race or religion or the country (or region) in which you were raised?

- Do you follow a political party because it is the one your parents supported?

- Do you think that having a detailed business plan, a physical office, and a lot of money are essential to the successful launch of a business?

- Do you root for a sports team because of where you live or because it is the one your parents rooted for?

- Do you hold onto a philosophical view of the world because it fits who you were twenty years ago?

- Do you assume that the definitions of words like "conservative" or "liberal" continue to mean what they did decades ago?

- Is your definition of a "good time" the same now as it was fifteen years ago?

- Have you ever declined to try something new because you "know what you like"?

Until the end of the 19th century, personal transportation was defined by some form of horsepower. The horse. The horse and carriage. The stagecoach. Then came the internal combustion engine and the rapid scale up of the automobile industry.

5th AVE NYC
1900

Where is
the
car?

Library of Congress

5th AVE NYC
1913

Where is
the
horse?

Library of Congress

The first photograph was taken of 5th Avenue in New York on Easter morning in 1900. It shows this great boulevard full of horse-drawn carriages. There is only one car. Thirteen years later, same place, Easter Sunday, there is one horse. This was more than 100 years ago when the speed of change was somewhere between $1/10^{th}$ and $1/100^{th}$ of what it is today. It illustrates how quickly our realities can change... how quickly a worldview of the horse as the dominant mode of transportation was replaced by the internal combustion engine.

If you were a Manhattanite in 1900, what was your most prevalent environmental problem? Yes, horse manure! What did the carriage drivers do to solve the problem? They attached large canvas bags to the rear end of the horses and at the end of their shift they dumped the manure (often into the rivers, which at the time were thought of as trash disposal sites).

What the Manhattanites did not see was that the solution to the horse manure problem was a new drive train.

The second point to draw from these two photographs is that often the solution to a present-day problem is something yet to be invented, a new reality often catalyzed by a new technology. This is important for us to fully understand. During the 2020s, humanity will face some massive problems, some that seem overwhelming given current mindsets, technologies and political constraints. As the internal combustion engine solved the horse manure problem (and created all kinds of new problems), new technologies and inventions will address the massive challenges rushing toward us.

Waves of creative destruction and unforeseen disruptions will largely alter the landscape of civilization this decade.

Another example of legacy thinking in the automobile industry in America (rightfully defined as a car culture), is what happened to the country in the 1970s. It was an axiom that Americans wanted big cars and wanted a new one every few years. Planned obsolescence was integral to the marketing of automobiles. Every fall the new cars would be unveiled at dealerships across the country. Today we have the iPhone 1 through 12. Our tech gadgets are the current form of planned obsolescence, though now more driven by tech innovations than design details such as tail fins. This of course will have to change in the 2020s because planned obsolescence is completely at odds with our critical need to reduce consumption. [More on that in a future book in this series.]

Then all of a sudden the game changed.

The Organization of the Petroleum Exporting Countries (OPEC) quadrupled the price of oil. Japan quickly reacted by producing small, inexpensive, unsexy but high miles per gallon cars. U.S politicians and union members responded by taking sledgehammers to imported small cars in publicity stunts. They set up tariffs to protect an industry that was already out of touch with global trends and consumer preferences (a real thick-headed example of holding onto legacy thinking).

Instead of trying to win by adapting, American politicians protected the status quo that had been triumphant in the 1950s and 1960s. As a young boy growing up in Chicago it was an incredible sight to ride with my parents through Gary, Indiana and see the world center of steel production belching fire as thousands of tons of steel were created for the auto industry. As an adult I drove this same route and saw only a toxic wasteland. The death of this industry occurred as humanity moved from the Industrial Age to the Information Age.

Legacies from this transition still scar the American landscape and political arena. The disintermediation that the Internet wrought over the last 25 years is an even larger manifestation of the collapse of legacy thinking. We now buy things online that would have been unthinkable 15 years ago. We now routinely make all kinds of transactions online with people and organizations all over the world. [More on this in the next chapter.]

We are in danger of limiting potential solutions to today's problems with legacy thinking... outmoded belief systems concerning what is possible, particularly in the area of technology. A clear example is the well-intentioned effort to teach young girls how to code. This stems from the current reality that coding jobs are in demand, well-paid, and overwhelmingly held by males. It's an attempt to address gender inequality. The problem is that teaching fifth grade girls how to code is a failed effort long term. By the time these girls graduate from high school, there will be no coding jobs. Computers are already self-coding. It will be a useful skill, like touch-typing, but not a means for earning a living.

Addressing our current and near-term problems using legacy thinking is largely a dead-end effort. The ideas, structures and concepts put in place during the 20th century are legacy thinking destined to collapse. Humanity entered the 21st Century with 20th Century constructs. Those belief systems are currently being replaced by new 21st Century ones.

We can illustrate this by returning to the internal combustion engine [ICE]. It was brought to market in the early part of the last century. Then it became institutionalized as our default mode of transportation. Now, in the 2020s, we are rapidly moving to transportation powered by electricity, soon to be augmented with hydrogen fuel cells. This will cause huge cascading disruptions. As we move to electric power in cars and public transportation, the sales and distribution model for ICE vehicles will collapse. Currently the largest profit center at a car dealership is the service department. ICE cars have thousands of parts. Electric cars have less than a few hundred. This means that the move to electric will drive traditional car dealerships out of business in the second half of the 2020s.

As stated in the prior chapter, COVID-19 is preparation for the rest of the decade in terms of accelerating the collapse of 20th century physical distribution models. The virus forced the shutdown of the movie theater business model. With no one going to the movies, there was no revenue. The pandemic supercharged the streaming model for entertainment. Then the studios reoriented their business models toward streaming. The number of theaters in the U.S. at the end of 2019 was approximately 40,000. By the end of 2021 that number will be less than 20,000.

Another example is higher education. The university model of teaching is centuries old. A teacher lectures to a room full of students and assigns reading from a textbook. Retention of the material is assessed via some sort of testing. Yet, the cost of higher education exploded in the last third of the 20th century. Higher education costs in the United States have increased more relative to inflation than any other expense.[1]

21st century technologies are disrupting this overpriced, over-valued business model and teaching paradigm. Online education is increasing exponentially. Confining higher education to people in their late teens and early twenties is also being disrupted. Life-long learning will become the norm as we evolve from the Knowledge Economy of the Information Age to the Learning Economy of the Shift Age. Lifelong education and continuing education will become the dominant model as the need for rapidly transforming skill sets becomes the norm.

A quote perfect for the current state of education, and for all aspects of dealing with legacy thinking, is from one of my intellectual heroes, Dr. Alvin Toffler. When asked in 2006 what the future of education might look like in the 21st century, he said:

"The illiterate of the 21st Century will not be those who cannot read or write, but those who cannot learn, unlearn and relearn."

In the 2020s, every person, every leader, every organized entity must elevate "unlearning" into an active verb.

Many of our conceptual constructs, business models and structures of physical distribution of the last two centuries will collapse as we move to more personalized technological forms of distribution.

This holds true at the highest level of organization. Nation states, states and provinces, and most certainly cities all will have to change how they handle numerous policies, taxation models, and infrastructures. We currently live in an infrastructure reality largely wrought in the 20th century. Policies that shape us are decades old and will not survive the realities of the 21st Century.

A way to view this clearly is to examine the two realities of the Shift Age.

---

[1] https://research.collegeboard.org/pdf/trends-college-pricing-2019-full-report.pdf

Chapter Four – Managing the Two Realities of the Shift Age

The two realities of the Shift Age, and of the 2020s decade, will cause massive cognitive dissonance. The move from a place-based culture, society and economy to a space-based one (where the importance of location is greatly diminished) has thrown hundreds of millions of people and millions of businesses into extreme cognitive dissonance. Let's take a look at the underlying dynamics that have triggered this.

The two realities of the Shift Age are the physical reality and the screen reality.

The physical reality is where humanity has always lived. In this century we also reside in a rapidly developing screen reality. The screen reality started to disrupt the physical reality in the 2010-2019 Transformation Decade. Now in the 2020s this disruption is rapidly accelerating.

In the early part of the 2010-2019 decade, humanity effectively reached cellphone ubiquity. The total number of cell phone subscriptions globally in 2000 was 740 million when the population was six billion. So 12% of the people on earth had a cell phone. In 2010 it was 5.3 billion subscriptions when the population was seven billion or 76% market penetration. In 2019 it was 8.3 billion with a population of 7.7 billion for a ratio of 108% (a percentage of population has more than one subscription). The number of total mobile devices is 14.2 billion in 2020 and will be close to 20 billion in 2025. This is more than double the global population. The growth in developing countries, where it is often the first stage of communications infrastructure, has been much greater than in developed countries.

Year	Total world population (mid-year figures)	Ten-year growth rate (%)
1950	2,556,000,053	18.9%
1960	3,039,451,023	22.0
1970	3,706,618,163	20.2
1980	4,453,831,714	18.5
1990	5,278,639,789	15.2
2000	6,082,966,429	12.6
2010	6,956,823,603	10.7
2020	7,794,798,739	8.7

2030	8,548,487,000	7.3

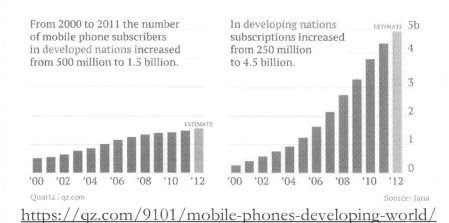

From 2000 to 2011 the number of mobile phone subscribers in developed nations increased from 500 million to 1.5 billion.

In developing nations subscriptions increased from 250 million to 4.5 billion.

Quartz | qz.com

Source: Jana

https://qz.com/9101/mobile-phones-developing-world/

Mobile Device Forecast (in billions) 2020 to 2024

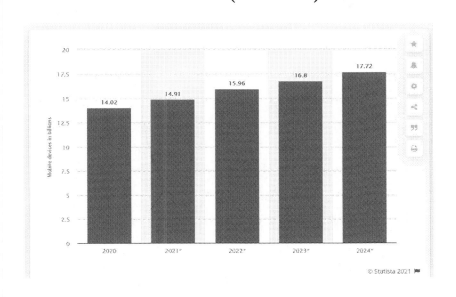

© Statista 2021

What this means is there is no longer any time, distance or space limiting human communication. This statement could not have been made 15 years ago. It was only 170 years ago that the telegraph came into wide use. This means that for the roughly 170,000 years that modern humans lived on the planet, only 1/1000th of that time have we been able to communicate without being face to face. The telegraph was quickly followed by the radio, the telephone, television, cable TV, and the Internet. More people listened to Mozart's music today than ever listened to his music live… because you had to be in the room!

The growth of smart phones has dramatically increased the number of photos being taken. In 2000 there were 86 billion photos taken globally. In 2020 there were 1.4 trillion photos taken.

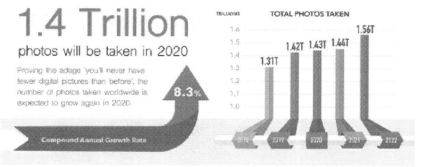

In the U.S. in 2020 it is estimated that 49.3% of all online sales, were made on mobile devices. By mid-decade this number could easily top 70%.

U.S. ECOMMERCE VS. TOTAL RETAIL* SALES

In $billions, 2018-2020

$523 B	$598 B	$861 B
2018	**2019**	**2020**
$3,626 B	$3,780 B	$4040 B

*Total retail figures exclude sales of items not normally purchased online such as spending at restaurants, bars, automobile dealers, gas stations and fuel dealers
Source: Digital Commerce 360 analysis of U.S. Department of Commerce data
Updated January 2021

Layer on top of this the fact that current state of the art smart phones are far more powerful than the super computers of the 1970s and 1980s, and we know that we have the greatest amount of distributed computing power in history......all in the palms of our hands.[1]

Our physical reality is giving way to a screen reality. This trend will continue throughout this decade.

I remember a day back in 1997 when I first plugged my laptop into a T-1 ethernet cable. (Remember, this was the time of "you've got mail" with dial-up connectivity.) All of a sudden, I was surfing the Internet. After about 30 minutes, I had the aha moment that almost everything that exists in the physical world would be recreated for an online screen reality.

This is the context for what has happened in 2020 relative to COVID-19. The virus accelerated the move from physical to digital, from in-person to online, and from physical reality to screen reality. We quickly migrated to working, attending school, shopping and being entertained on digital platforms. For example, at year-end 2019, Zoom had 10 million daily users. By early June 2020, it had 300 million daily participants, a 30-fold increase in six months. Our meetings went to the screen reality, our buying went to the screen reality, my presentations as a futurist went from the physical stage reality to a live streaming green-screen reality.

A profound effect of this migration is that it put a spotlight on the greenhouse gas emissions that are a byproduct of physical reality.

Dual Realities to Manage

Here is a completely different view of the two realities, one clearly demarcated by generations. If you are what I call the Digital Native Generation (often referred to as Generation Z) born since 1998, you have spent your entire conscious life in the screen reality. This generation is the first to matriculate K-12 and high school managing two identities, the one they live in the physical reality and the one they live in the screen reality. They have one foot in each world, and must manage each differently.

If you are a late Millennial born in the 1990s, you have come of age with these two realities as well. If you are a Boomer or GenXers, you are digital immigrants as you have come to the screen reality as an adult. This is why these older generations turn to their children or grandchildren for tech support.

I have a personal view of media technologies and generations that is difficult to document but seems anecdotally true. The media that was dominant in one's life up to the age of 21 is the default technology for that person for the rest of their lives. My father and mother, born more than a century ago, grew up with radio and newspapers. All their lives they read the daily newspaper and had the radio on at some point during the day or evening. As a boomer, I was of the first television generation, where the shared popular culture was TV. To this day, most Boomers default to TV. Millennials grew up with computers and laptops and default to those with mobile a close second. Digital Natives are mobile oriented and the latest generation, Generation Alpha, is totally mobile in its orientation.

Place to Space

This will be more deeply explored in a future book in this series. Not only do we have the two realities to manage, but our screen reality is also connecting us spatially. Prior to the telegraph, we had to be in the same place to communicate in real time.

COVID-19 has accelerated this dynamic. Think of higher education. The whole expensive package that is called the

American college experience has been decimated. Colleges can no longer profit from room and board and many other services provided for a fee on campus. No longer can they charge the same for online and in-person courses. Being on campus is obviously place-based. Being online is space-based.

Colleges that have places in their names now have thousands of students that don't live anywhere near the institutions. Both Arizona State University and the University of Southern New Hampshire, two of the most successful online college programs, are place-based and yet thousands of their students never go to campus. USNH has 3,000 students on campus and 135,000 students online. ASU had a record fall enrollment in 2020, when most colleges and universities saw enrollments tank. Almost half of ASU's students are online only.[2]

The move to streaming in the home increased dramatically during 2020. The number of subscribers to digital video streaming services such as Netflix, Amazon Prime, HBO Max, Disney+ and Hulu grew exponentially between March and September. At the same time most indoor movie theaters closed for months. Again, the place-based distribution model established in the 20th century was permanently curtailed by the pandemic. This migration was already going to happen during this decade, but COVID-19 accelerated the change from several years to just a few months.[3]

New habits have been formed during the pandemic. What this means is that what we thought of as normal is not going to come bouncing back. Once everyone has been vaccinated, many people will go back to movie theaters, but a sizable number will have embraced the lower cost and increased convenience of watching movies in their screen reality whenever and wherever they choose.

Working from home is a new screen reality. Many millions of people no longer work in offices. They are no longer centralized, but distributed. The concept of place has been supplanted by the work space via screen reality.

In 2012 I gave a TEDx talk called "The Concept of Place Has Changed Forever."

In the 2020s, our screen reality will grow ever more dominant relative to our physical reality. Work, entertainment, education, telehealth and shopping will all continue to migrate away from place and physical reality to space and screen reality. This is a sure trajectory for the rest of the decade.

[1] https://www.techrepublic.com/blog/classics-rock/the-80s-supercomputer-thats-sitting-in-your-lap/

[2] https://www.insidehighered.com/digital-learning/article/2019/12/17/colleges-and-universities-most-online-students-2018

[3] https://www.theverge.com/2020/6/2/21277006/zoom-q1-2021-earnings-coronavirus-pandemic-work-from-home

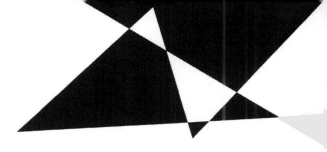

Chapter Five - Past Eras of Massive Change

The decade of the 2020s will be the most disruptive decade in history. This is largely due to the amount and speed of change in these ten years. There have been times in the past when truly transformative change occurred over several decades. Over a 50-year period society and economies can make such massive changes that those alive at the end of the period can hardly imagine how their grandparents and great grandparents lived.

Another reason for this degree of disruption today is our degree of connectedness. New ideas, technologies, cultural fads, etc. are proliferated globally at the speed of thought. Past periods of change took place more slowly hampered by less robust communications and geographic and political barriers.

Let's take a look at some of the transformations in history as reference points to what we face by the 2030s.

1450 –1525 in Europe

The Internet has been the single greatest agent of disintermediation in history. As it became ever more available around the world in the last 25 years, it has disrupted a majority of business models and created new ones that have redefined how we live.

Second to the Internet as a force of disintermediation was the invention of the moveable type press by Johannes Gutenberg in the 1440s. He was the first person to print the Bible in 1455. This invention disintermediated the Church's control of knowledge… making it available to the aristocracy (and then the middle class). This new way to inexpensively mass-produce books transformed the world. Books on a variety of topics brought revolutionary ideas and priceless ancient knowledge to literate Europeans, whose numbers doubled every century. Knowledge was now accessible to anyone who could read.

In addition to books, the moveable type press enabled newspapers to spread news of the world (and propaganda). Knowledge became broadly disseminated. People could learn what was happening beyond their immediate locale. It also distributed literacy by making the skill useful.

Due to the printing press, when Martin Luther posted his "95 Theses" creating an alternative to the Catholic Church in Germany, the document became readily available in England and elsewhere in Europe within a short period of time. Knowledge, news, historical events could now be disseminated widely, greatly increasing knowledge of the contemporary world. The printing press began the era of mass communications that we live in. When writing or speaking about mass communications, one should turn first to Marshall McLuhan:

"With Gutenberg, Europe entered the technological phase of progress, when change itself becomes the archetypal norm of social life."

And

"The invention of typography confirmed and extended the new visual stress of applied knowledge, providing the first uniformly repeatable commodity, the first assembly-line, and the first mass production."

And

"The portability of the book, like that of the easel-painting, added much to the new culture of individualism."

So this 75 year period started with Gutenberg's invention which democratized knowledge, increased literacy, spread ancient and contemporary ideas, initiated mass production and the culture of individualism. All huge changes from the Middle Ages.

Also in this transformative period, Martin Luther started the Protestant Reformation, the Renaissance exploded in Italy, America was "discovered", and Arabic numerals were adopted in Europe. A woman alive in 1525 was hard pressed to imagine the daily lives of her grandparents.

1775 – 1825

Now we move from a 75-year period of transformation to a 50-year one, with even more change in less time. Capitalism, Democracy, and the Industrial Revolution all began during this era. Here is just a short list of significant ideas, events and inventions that occurred during this time:

- Adam Smith published "Wealth of Nations" creating capitalism.

- James Watt invented the steam engine, launching the Industrial Age.

- The French and American revolutions created democracy.

- Eli Whitney invented the cotton gin… automating a part of the clothing industry.

- Edward Jenner invented the smallpox vaccine… the first ever vaccine.

- Friedrich Surturner discovered morphine, enabling medical surgeries.

- Charles Babbage built the first programmable computer.[1]

It was a rich time for the realm of culture. Here are just some of the notable creators in literature and music:[2]

Austin,
Beethoven,
Blake,
Brothers Grimm,
Coleridge,
Cooper,
Gibbon,
Goethe,

Hayden,

Heine,

Kant,

Mozart,

Paine,

Scott,

Shelley,

Voltaire.

Geo-Political Events of this time transformed the world:

- The American Revolution

- The French Revolution

- The Louisiana Purchase

- Napoleon becomes emperor

- The British-American War

So this 50 year period was the beginning of what became the Industrial Age, Capitalism, Democracy, the nation state, deeper and richer cultures in Europe and the US. Again, someone in the 1820s could look back at the lives of her grandparents and be astonished at how limited and mostly rural their lives had been. Looking at the inverse, those grandparents would be incapable of understanding the society, culture and economy of their granddaughter.

1875 – 1925

This truly was a transformational and disruptive time. It started right after the Industrial Age North defeated the Agricultural Age South in the American Civil War. That event was the final curtain for the Agricultural Age in America and Europe.

The list of disruptive inventions for this 50-year period is truly amazing:

- Alexander Graham Bell patents the telephone.

- Edouard Benedictus invents laminated glass.

- Guglielmo Marconi invents what became the radio.

- Hans von Pechmann synthesizes polyethylene, the most common plastic in the world.

- Heinrich Hertz publishes conclusive proof of electromagnetic theory and also demonstrates the existence of radio waves.

- Huber Cecil Booth and David T. Kenney each independently patent the vacuum cleaner.

- Jacques Brandenberger invents cellophane.

- James Blyth invents the first wind turbine used for electricity.

- John Kemp Starley invents the modern bicycle.

- John Loud invents the ball point pen.

- Joseph Swan and Edison patent the incandescent light bulb.

- Karl Benz invents the first gas powered automobile.

- Nicola Tesla developed alternating current electricity.

- Rudolf Diesel invents the diesel engine.

- Sir Charles Parsons invents the modern steam turbine.

- The Wright Brothers make the first fixed-wing, motorized aircraft.

- Theodore Kober invents the first Zeppelin.

- Thomas Edison invents the phonograph.

- Waldemar Jungner invents the rechargeable nickel-cadmium battery, the nickel-iron electric storage battery and the rechargeable alkaline silver-cadmium battery.

- Whitcomb Judson invents the zipper.

- Wilhelm Rontgen discovers X-rays.[3]

In addition, during this 50-year period, the use of elevators and indoor plumbing became widespread. Critical developments relative to the rapidly growing cities.

This time began with the Agricultural age in the developed countries of that time. And by the end all were fully in the Industrial Age. This 50-year period delivered more change in how people lived than any prior 100 year period.

"The Industrial Revolution caused a centuries-long shift in power to the West; globalization is now shifting the balance again." – Dennis C. Blair

At the start of this era, 70% of the U.S. population lived in rural areas. By the end of the it, 60% lived in urban areas. This dynamic completely changed economics, society and culture. The numbers for Europe and elsewhere were similar.

In addition, Industrial Age Capitalism exploded with the Gilded Age, Robber Barons and the greatest accumulation of wealth to that date. The names of the age, Carnegie, Rockefeller, Gould, Ford, Frick created the wealthy families whose names are influential to this day.

Someone living in a big city in 1925 in Europe or the U.S. would find stories of how their grandparents lived in the 1860s and 1870s to be almost beyond comprehension. Conversely, imagine those grandparents being told of telephones, radios, cars and airplanes!

"To join in the industrial revolution, you needed to open a factory; in the Internet revolution, you need to open a laptop." – Alexis Ohanian

I sit writing this in Florida in 2021. My grandparents emigrated from Canada to Sarasota, Florida in 1900. They lived in Canada and in Florida with horse and buggies, limited electricity, no telephones or radios. That was just two generations ago. The speed of change has been accelerating since 1875 and it continues into the 2020s.

Now let's put these three transformative times in the perspective of the 2020s.

The amount of change that occurred 500 years ago over a 75-year period will occur now during this decade. The amount of change from 1775 to1825 will occur in the 2020s and at least 75% of the amount of change from 1875 to1925 will happen this decade. Unprecedented change and disruption.

The first two eras happened over several generations as people did not live as long and communication over distance happened at the speed of horse. Therefore, minimal cognitive dissonance. There were the game changing inventions of the printing press and later, the steam engine, but only small segments of the global population experienced these inventions during each of these eras. The third era was the first time such profound changes might be experienced within a single lifetime, but that was dependent on where one lived. Even in the last of the three eras, the amount of change was experienced over a lifetime, not in a small percentage of that life.

Third, it was only during the last 10 years of the 1875-1925 era, when radio reached a sizable number of homes and people could learn about events outside their physical reality. Compare that to today when the information is available to all at the speed of light via fiber optics. Ever growing electronic connectedness is an accelerant for cognitive dissonance unprecedented in human history. Immediacy of information puts us in a constant state of assimilation and digestion of data. A constant stream of data that must be received, understood and placed into our constantly changing sense of "reality".

The first book in this series was setting the stage for all that will be in the 2020s. When writing that book I knew that this one about cognitive dissonance had to be the second book of the series as it addresses an underlying emotional, intellectual and psychological disruption we will all need to deal with, and it underscores all the issues to be discussed in the remainder of the series.

Understanding cognitive dissonance and learning how to live with it, is an essential requirement for successfully navigating the 2020s.

[1] https://en.wikipedia.org/wiki/Timeline_of_historic_inventions/

[2] https://cmed.ku.edu/private/classical.html/
https://en.wikipedia.org/wiki/List_of_years_in_literature

[3] https://en.wikipedia.org/wiki/Timeline_of_historic_inventions/

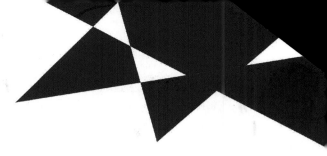

Chapter Six - Learning to Live in a State of Change

"Standing still is the fastest way of moving backwards in a rapidly changing world." – Lauren Bacall

The only constant in the Universe is change. Nothing is fixed. All is change. If there was no change, there would be no time. Time is a metric of change. Through time things change. Humans may not be able to discern the changes that take millennia to occur, but that doesn't mean that change is not occurring.

Change is a fundamental reality that we cannot live outside of. That is one of the reasons that we rely so heavily on our habits. Habits create certainty and reliably deliver the same experiences over and over. This is how I get ready in the morning. This is the route I drive to work. This is the place I always get my coffee. This is when I post to social media. This is how I get my exercise. All this activity buffers the reality that change is constant. We try to establish certainty, so we have a way to check out from constantly navigating change.

The Shift Age, the first third of the 21st Century, is the time when the speed of change has accelerated so much, it is no longer is a linear progression, but has become environmental. We live in an environment of change, where multiple aspects of life, of our individual and social lives, are changing simultaneously.

All change, everywhere, all the time. Cognitive dissonance.

In the 2020s, change will continue to accelerate. What you are doing in 2020 will not be what you are doing in 2025, let alone 2030. How you are doing what you do will change. What you think is reality will change. How and where you work will change. How you communicate will change. The energy sources you use will change. Your form of transportation will change. How you view consumption and what you consume will change. Your relationship with government will change. Every aspect of your life (and mine) will change in the 2020s.

Entering the 2020s decade, I feel that as a futurist I am the canary in the mine [see chapter 3 of Book 1]. This means that while I will continue to look into the future of the 2030s, 2040s and beyond, it is this decade where the trajectory of humanity will be determined. As I wrote:

"We are at that inflection point, the nexus of change, the historical moment which will set civilization on its path for the remainder of the 21st century. The 2020s will define the direction for humanity, including our survival as a species."

Although fully present in the NOW, as a futurist, I see the future as it arrives today. This is not how most people experience the present. Let's take a look at this real and immediate phenomenon.

This chart [also in Book 1] shows that in the last 125 years the speed of technological adoption into the broader population has dramatically accelerated in the Shift Age and the 21st Century:

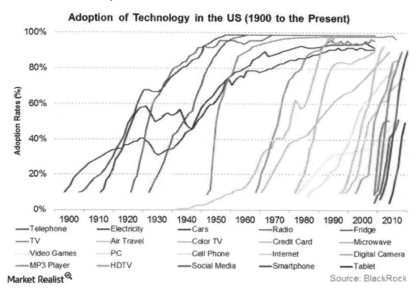

In this chart we see how we all are spending more time in the Screen Reality:

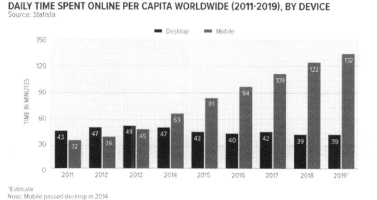

Time spent online tripled over the last decade. This will continue to increase in the 2020s. Mobile is the dominant technology of the 2020s.

What the future of social media might be in the 2020s is now a bit up in the air. Governments around the world are beginning to challenge the powerful tech platforms. Yet, if last decade is any barometer, our collective engagement will quadruple at the minimum.

NUMBER OF PEOPLE USING SOCIAL MEDIA PLATFORMS
Source: Statista and TNW (2019)

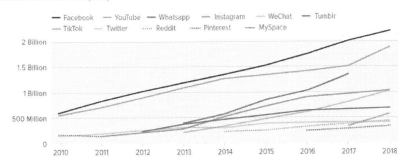

Note: Estimates correspond to monthly active users (MAUs), Facebook, for example, measures MAUs as users that have logged in during the past 30 days.[4]

Our modes of transportation will dramatically evolve in the 2020s. Looking back to the last decade, electric vehicle sales increased by 14,000% in the US. The rate will continue to increase in this decade. Prices for batteries will continue a steep decline. All major car manufacturers have doubled down on EV production. There will be an insatiable demand for batteries and ever lower costs will trigger a highly competitive market, resulting in technological breakthroughs that lower costs even more.

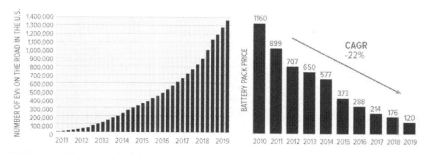

ELECTRIC VEHICLES ON THE ROAD IN THE US & BATTERY PACK PRICE
Source: Global X Research, Benchmark Mineral Intelligence, Edison Electric Institute, BloombergNEF

CAGR (Compounded Annual Growth Rate)
Note: From 2010 to 2018 numbers reflect real $/kwh per BloombergNEF estimates and 2019 reflects Benchmark Mineral Intelligence estimates.

The phrase "information explosion" became popular during the last third of the 20th century. That term was relative to the prior 100 years. That information explosion looks pitiful compared to where we are now. Note: a zettabyte is equal to a trillion gigabytes.

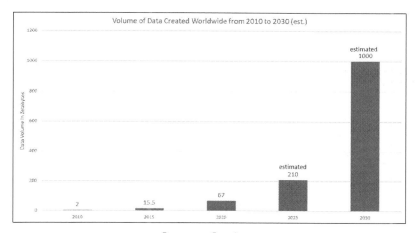

Source: Statista

Do you think that a 15x increase in data/information in 10 years might create some cognitive dissonance?

All this data will be made available to us via a tsunami of connected devices:

GLOBAL NUMBER OF CONNECTED DEVICES BY 2030 (IN BILLIONS)
Source: Global X Research, Ericsson Mobility Report, Cisco

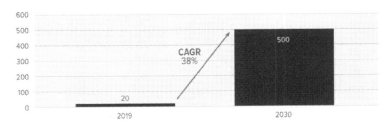

Our everyday degree of connectivity and information consumption will be overwhelming. We will all become more reliant on Technological Intelligence [TI] simply to get through a day.

[Note to reader: I choose to use the label Technological Intelligence (TI) rather than Artificial Intelligence (AI). I know I am swimming upstream, but I think that using the term AI is a mistake. In 2015-16, TI was beginning to look like a "hockey stick" growth curve. I looked up the word "Intelligence" in five different dictionaries. Not one of them used the word "human" as part of the definition. Whales are intelligent. Dolphins are intelligent. The Universe is intelligent. Why do we think that "intelligence" means human only? Because we look at the universe entirely through an anthropocentric lens. The reasoning is that if it is not human, it must be artificial. Wrong.

The discussion around AI has a bias of negativity. On some subconscious level, the word 'artificial' means less than or fake. The term AI has been exclusively to describe the realm of technology that can learn on its own. So it is intelligence in technology or of technology, hence TI.

This anthropocentric view is also a primary cause of our climate crisis. The majority of humanity has determined that our species is above all else in the natural world. The natural world is for us. We are not of the natural world. If, instead, we viewed ourselves as part OF the natural world, that we are ONE of many species on earth, we would behave accordingly. Then our actions would be taken only after considering the consequences on all other species, plants and animals, had been considered. So I choose to use the term Technological Intelligence for this book and the rest of the series of books about the 2020s.]

Technological Intelligence

Technological Intelligence will cause massive change and reorient how we do things… including our work. It will eliminate many jobs. Losing a job triggers upset, professions losing double-digit percentages of jobs will cause cognitive dissonance. The legal and accounting professions will be eviscerated by TI in the 2020s. Estimates range from 25 to 50% of all jobs in these two white collar professions will be gone by 2030. The first category of jobs to go will be ones that have a high degree of repetitive actions. Accountants do vast amounts of repetitive work such as tax returns, monthly financial reports, and certified transactions. Lawyers do high volumes of repetitive work as real estate transactions, wills, incorporations, all kinds of things that have only a small amount of variance. So, white collar professions with high percentages of repetition will be lost to technology.

Of course there will be an increase in manufacturing jobs in the 2020s, but the vast majority of them will be filled by robots:

NUMBER OF INDUSTRIAL ROBOTS WORLDWIDE
Source: Global X Research, Oxford Economics, International Federation of Robotics

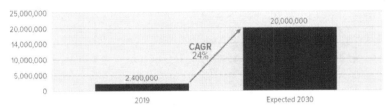

CAGR (Compounded Annual Growth Rate)

TI will replace between 10 and 40% of white collar jobs depending on the country and the nature of the work.

What to do?

Either become comfortable living in full blown cognitive dissonance, or choose to drop out of society, business and active participation in the world.

This will be disorienting for most of us. Some people see change as an opportunity, but many more are uncomfortable with change. Certainty is valued as an anchor in life and business. The more one tries to hold on to this viewpoint, the more cognitive dissonance one will feel.

- Childhood is punctuated by basic and measured steps of newness. Each year of school marks progress. We run a gauntlet of social, educational, physical changes that we are called upon to matriculate.

- Puberty is a physical metamorphosis into a mature being.

Then we become adults and time accelerates. Turning 25 is a four percent increase from 24. The percentage of newness usually decreases each year as we have more life experiences under our belts. We get married 'for better or worse until death do us part'. We lock into our lives, relationships and careers. Rarely does each year bring the amount of change we experienced growing up.

Why is this relevant to the 2020s and cognitive dissonance?

The amount of change ahead for each of us in the 2020s will be a bit like childhood in the sense that each year of this decade will be much different than the prior year. We will be back into double digit amounts of change per year, as we were in our youth.

In Chapter 2 the comparison was made between COVID-19 and learning to ride a bike with training wheels. Think about the passage of time in 2020. At times it felt as though time was slowing down. At other times we were amazed that so much time had passed. Our relationship with time became disorienting. It was no longer Monday, Tuesday, Wednesday, Thursday, Friday, Saturday and Sunday. It was today, today, today, today, today, today, and today. There was a sameness, and there were also days that were unique because something happened (someone we knew fell sick, a business we frequented closed their doors, something crazy happened in the political realm). Weekdays and weekends were not differentiated in the way they used to be. Lockdown meant rarely leaving the house… no matter the day.

Our externalities became less personal and more collective. We didn't socialize outside the home, but we watched more television. We had a greater sense of our collectiveness due to the common experience we were all going through. The killing of George Floyd became much larger than prior such incidents because the whole world was watching.

The Screen Reality will continue to increase in our lives. Ever more commerce, communications, conferences, meetings, viewing, listening and socializing will be in the screen reality. This domination of the digital and virtual will be disruptive to our physical lives and sense of self. The cognitive dissonance between the physical and screen realities will increase. Many will largely default to one or the other. Those that deepen their sense of place and identity related to their specific, physical reality will risk significant cognitive dissonance whenever they do choose to jack-in to the data explosive screen reality.

"If somebody describes the world of the mid-twenty-first century to you and it doesn't sound like science fiction, it is certainly false. We cannot be sure of the specifics; change itself is the only certainty." – Yuval Noah Harari

The converse will also be true. Those that get lost in the screen reality will have unpleasant and difficult experiences in the physical reality. The ongoing truth will be that each and every one of us will experience wide swings between the two, even on a daily basis. That will be a fundamental trigger of cognitive dissonance, the need to manage two realities all the time.

Who one is, what one does for work, what one does for play, where one lives, how one describes relationships will all affect this balance. Someone living out in the country will still want connectivity. Someone living in an always connected urban setting will want to get away or unplug. It is this constant back and forth between two realities that will trigger cognitive dissonance.

Think back to the three eras of massive change we looked at in the last chapter. There was only the physical reality. There were no screens. Until the 1920s, when radio became popular, there was no reality beyond what we experienced in the physical world.

A quote from the poet laureate of the Baby Boomer generation has always served me well when reflecting on day in and day out change:

"Those not busy being born are busy dying." – Bob Dylan

Being busy being born is a way to minimize cognitive dissonance.

Every year you choose to be involved in the world will require a new birth of a greater part of self than in the past. Some of us will actually change more in a year in the 2020s decade than we did over five years a few decades ago. Others will not, but will still feel swept up in the flows of newness that alter what happens in our lives on a daily basis.

The 2020s will trigger an unprecedented level of rebirth at all levels.

[1] https://www.elsevier.com/connect/medical-knowledge-doubles-every-few-months-how-can-clinicians-keep-up

Chapter Seven – The Big Things

In Book One of this series, I set forth the four overarching dynamics of the 2020s:

- The Age of Climate
- The Age of Intelligence
- The Reinvention of Capitalism and Democracy
- A New Emerging Consciousness

Most of the changes in this decade can be placed under one of these four dynamics. They represent the major constructs that we will have to matriculate. The degree to which we are forewarned we can anticipate. The amount of anticipation we can bring will help us to flow with the rapid changes. Anticipating these changes will also help us minimize the cognitive dissonance resulting from all the adaptations needed. Here are some challenges that everyone reading this book will have to meet to one degree or another.

Work and the workplace will evolve:

- What you do for a living will likely change.

- How you do your work will most certainly change.

- Past measurements of success will largely change. There will be entirely new metrics for success.

- Where you work will change.

- Your income and/or revenue will not derive solely from your work.

- The Age of Intelligence will dramatically alter how we work, collapse the workforce, and provide entirely new ways to work.

How we live and how we consume will change in significant ways:

- The Age of Climate will deliver a collective reality for facing this existential threat. That will force changes in how we live.

- We will view consumption differently. We will move to "Conscious Non-Consumption."[1] Conspicuous consumption will become conspicuous non-consumption.

- All forms of transportation will dramatically change.

- There will be a complete overhaul of our energy sources, delivery systems and usage.

- The Age of Intelligence will profoundly change how we live, how we interact with others, how we work and how we consume.

Economics and the Global Economy

- Humanity first (#humanityfirst) becomes a more dominant dynamic overshadowing even the transition to stakeholder value.

- The global wealth inequality gap will continue to grow. This threatens social and cultural stability. Nations will have to address it or risk populist uprisings.

- New tax structures and more comprehensive safety nets will be created.

- The last 30 years of debt underpinning the global economy will have to be reckoned with. There will be an unprecedented financial disruption.

- Digital banking, digital currency, digital investments and digital assets will trigger rapid and massive changes to banking systems.

- 20% of the global GDP will undergo profound disruption as the world moves from 75% of all energy coming from fossil fuels to 25 to 40% by 2030. This will have cascading consequences through the global economy. There will be explosive growth in all forms of clean energy. The oil and gas industry will first experience massive losses, and then new revenue as it finds its place in an anti-carbon world.

- Transportation is also 20% of the global economy. There will be more than just a transition from Internal Combustion Engine (ICE) vehicles to electric and hydrogen-powered vehicles. The Age of Intelligence will enable an ever-increasing number of autonomous/driverless vehicles. Electric airplanes with 500-mile ranges will be deployed.

- Wars between nation states over trade, resources and capital flows will evolve to more integrated global coordination and oversight.

- There will be a growing understanding, due in large part to our climate crisis, that Gandhi's quote reflects our new reality: *"There is a sufficiency in the world for man's need, but not for man's greed."*

Democracy

- During the early part of the 2020s, democracy will continue to be attacked around the world. Autocracy will be on a temporary ascendency in the first part of the decade, but will collapse in the second half.

- Democracy will seem to be less democratic and less accessible.

- Technology will flood into democracies to increase speed, transparency and efficiency. What we now do on our smartphones in health, investments and social media we will be doing relative to our participation in democracy.

- Universal broadband coverage will enable real time voting electronically.

Facing Our Climate Catastrophe

- How we live, where we live and in what accommodations we live will change for many.

- How we consume, what we consume and how much we consume will evolve.

- How we travel and what types of transport we use locally, nationally and globally will change.

- Creating a global authority for planetary policy will elevate discourse up from the nation state level. Climate is a global issue and must be addressed by all of humanity.

- To avoid long-term degradation of civilization, massive near-term changes in how humanity lives on spaceship earth will be mandated.

- The global workforce will continue to transition to work at home or distributed workplace models.

Education at All Levels

- K-12 education will diversify. Technology, including A/R and V/R, will be integrated into learning. The trend toward learner-centric customized instruction rather than grouping by age into grades will rapidly accelerate.

- As with everything else, there will be a move from delivery in a specific place to delivery in multiple places online and off. K-12 will change its funding model from simple local real estate taxes to other forms of tax revenue.

- The several centuries old model of college – a single teacher lecturing a room full of students – will continue, but will gradually be augmented or replaced by other teaching models.

- A fundamental flaw in the legacy model of higher ed is that the costs have risen faster than almost any other area in the last 50 years. Student loan debt and cost versus value will be addressed.

- In general, as it has almost everywhere else in society, technology will disintermediate old forms of delivery systems.

Medicine and Healthcare

- COVID-19 has, in many areas, condensed years of change into a single year. The pandemic has revealed the failures of practically every country's public health system. Few countries demonstrated preparedness for the virus.

- Countries that have strong national health systems, such as the U.K., Israel and the UAE have done better with vaccine distribution than the private for-profit U.S. system. The COVID catastrophic death event in the U.S. will trigger the development of an on-going "just in case" public health plan that stands in readiness for the next virus.

- By the end of the 2020s, the U.S. will replace its current, profit-focused health care system. Which will create another significant, yet positive, disruption.

- The major change agents in medicine will be Big Data, Technological Intelligence and DNA. Key treatments will be customized to individuals. Technological Intelligence already has a track record for more accurate diagnoses of early-stage disease than doctors.

- Digitization will transform health care. One example will be embedded chips with one's medical history (no more filling out forms on clipboards).

- Memory will be fully mapped in the brain this decade. One result will be a technological solution to memory loss via memory chips.

Entertainment

- COVID has collapsed years of change in distribution models and delivery systems into months. Streaming is the new default choice for viewing at home or on mobile.

- The physical distribution models of movies, perfected in the 20th century, will remain but in a much reduced capacity.

- Virtual Reality will finally take off with the next couple of VR goggle generations improved in size (smaller), functionality, price and with the embedding of sensory triggers.

- The Internet will increasingly become the backbone for all entertainment. With live-streaming via VR and AR technologies, streaming will begin to approximate the in-person experience.

Communications

- New smartphone hardware features and functions will slow. There will be some leaps ahead with holographic interactions and projection from our smart handheld devices.

- As 5G unfolds in the developed countries, it will bring live, real time, multi-streaming immediacy to tens of millions. This will deliver a major upgrade to personal and corporate communications.

- Embedded communication chips will enable choices for people wishing to be liberated from handheld devices.

- Brainwave computer interface will become mainstream by the middle of the decade, enabling the next step of human evolution by combining technological intelligence with human intelligence.

- There is a possibility for real-time uploading of thoughts, then memories, then consciousness to the cloud, where it can be retrieved and shared with others.

Family, Relationships, Offspring

- Legacy thinking around our concept of a nuclear family will collapse.

- The Millennial and Digital Native generations will have far fewer children than any prior generations.

- The binary idea of man/woman in sexual relationships will grow in gender fluidity.

- Diversity will grow in acceptance everywhere around the world.

- Multi-racial relationships will continue to increase in number and acceptance.

- The influence of organized religion on society will decrease as younger generations move away from religion.

Money and Finance

- Global indebtedness has already increased substantially with COVID-19. It will continue to increase until sometime in the mid-decade when a debt reset crisis will occur.

- Digital currencies and a distributed financial system will become a viable alternative to centralized finance and fiat currency.

- There will be micro crypto or digital currencies within cities, regions, and groups. These "mini-digital" currencies will be designed to solve local or regional problems, create closed-loop groups and even be earned for services as a new digital form of barter. Entire communities or transnational organizations will have discrete digital currencies with qualities unique to the group.

So in practically every sector of human activity, society and economics there will be significant change. Work, transportation, energy, money, work-life balance, concepts of family, human habitats, will all experience massive changes. For those not comprehending this disruptive time, a severe amount of cognitive dissonance is in store.

Now we turn to the duality of the realities and the self which underlies all of this creative destruction.

[1] https://finiteeartheconomy.com/

Chapter Eight – The New Dual Self

"In this electronic age we see ourselves being translated more and more into the form of information, moving toward the technological extension of consciousness." – Marshall McLuhan

The 2020s decade has ushered in the new dual self. The dual self is a distinctive characteristic of the Shift Age[1]. Chapter 4 was about managing the two realities of the Shift Age – the physical reality and the screen reality. As stated in that chapter:

"The physical reality is where humanity has always lived. In this century we also have the rapidly developing screen reality. The screen reality started to disrupt the physical reality in the 2010-2019 Transformation Decade and now in the 2020s this disruption is rapidly accelerating."

The COVID-19 virus has accelerated the move from physical reality to screen reality. The work from home phenomenon, the distributed workplace, is possible due to the connected screen reality put in place in the last decade. The virus accelerated the rate of change from several years into one.

Working from home is a screen reality. Working in an office is a physical reality. While many will move back to the physical office reality post-COVID, that percentage will be much lower than people expect. It might be less than 70% of office workers once the population is vaccinated.

Most people in the connected world toggle back and forth between their physical reality and their screen reality. It's a constant balance we all must manage.

This ties in with the overarching dynamic of the movement from Place to Space noted in Chapter 4. We can be anywhere in a connected space and communicate with anyone else (coworkers, family, strangers). We do not have to be in a particular place. The particular place is now the screen.

There is an historically unprecedented split developing in our sense of self. Starting with the generational lens, Boomers grew up with a place-based sense of self. The individual. The existential hero. Alienation of the self from the larger society or culture. The cowboy or the rebel on the motorcycle riding off into the sunset. Me.

Contrast that with the dominant descriptive of the Millennial generation. Dating or socializing in groups. Much more of a collective social sense. We.

Digital Natives were the first generation to grow up with social media, which is a shared, collective experience. Two selves were created, the in-person and the online person. A bifurcation of the developing sense of self.

Now fold in the duality of the screen and physical realities. Both must be perpetually managed. The two selves become real. This is me in physical place and this is me in cyberspace.

Millennials and Digital Natives (and now joining, the Alpha Generation, children of the Millennials) have more smoothly integrated these two realities than GenXers, and certainly the Boomers. These two older generations have an established self in place and largely regard the space self as something they use technology to enter. They see the smart phone as a tech device. The younger generations see it as an extension of who they are.

Given that the younger generations are in ascendancy, this sense of duality, the integrated self of the two realities is the emerging identity of the 2020s.

"Societies have always been shaped more by the nature of the media by which men communicate than by the content of the communication."
— Marshall McLuhan

Social media has been widely ridiculed and scorned due to an abundance of negative, mindless and snarky content. We read the posts and judge. McLuhan's quote is how I view social media. It is not about the content itself, but the fact that billions of people are sharing content. It is the ability to share that is the power of social media. To go back to a Gutenberg consciousness, we all learn to write, but that doesn't mean that what we write is important. We all connect with many others as we share. Connected sharing, at the speed of light.

The Next Step in Our Collective Self

"We shape our tools and afterwards our tools shape us." — Marshall McLuhan

I find this quote profound in the largest sense. What we create ultimately shapes us. Just think how ICE vehicles completely shaped how Americans live post WWII. We are a car culture. We created computers, then personal computers and now they shape us. We created the smart phone and now we see how much it shapes our lives, our culture and our sense of self. It has become our default device. Most of us interact with our phones more than any other device.

The Accelerating Connectedness of the Shift Age referenced in Book 1 is still the most dynamic of forces at play in the world today. This force, and the devices that connect us, have created a new reality I call "the Neurosphere". This is a pulsating, synaptic, Internet-based technological model of what a new consciousness might look like.

As I wrote in my 2013 book "Entering the Shift Age"[2]:

"This connectedness, happening at the speed of light via fiber optics, is creating an entire new place: the Neurosphere. Our physical reality exists in the biosphere — the thin surface of the planet where life exists. But this new, rapidly growing Neurosphere is the electronic extension of our collective neurological activity. It is a pulsing cyber-repository of humanity's creative brainpower, its knowledge, history, culture, social interactions, entertainment and commerce. This is a global village vastly more comprehensive and interconnected than Marshall

McLuhan could ever have envisioned when he coined the phrase "global village" more than forty years ago. We now live in a two-reality world: the physical reality in which we live and the Neurosphere reality of the screen that connects us to everything and everyone else on the planet."

The Neurosphere is now much faster, more connected and much more based on video than when I wrote those words eight years ago. In the 2020s, the Neurosphere will move us ever closer to lift-off into a preliminary collective consciousness. This new global consciousness is one of the four overarching forces that will shape the 2020s. The Neurosphere, as it exists today in 2021, will be the technological departure point for this new phenomenon.

At the end of "Entering the Shift Age", the last chapter was a theoretical look back from the vantage point of 2033, 20 years after the book had been published. The book ended this way:

"We now see the logical merging of physical and screen realities. How could we live without it now? Finally the evolutionary shift in consciousness predicted in that first edition of "Entering the Shift Age" has turned out to be so much more beautiful and meaningful than could have been imagined. The reality of the collective consciousness and awareness most of us regularly experience seems even greater than the dreams and early visions of it going back a century. It really is becoming an evolution shift even greater than imagined. In and of itself it has altered humanity more than anything else these past twenty years. How wonderful to be alive now!"

When I wrote that almost 10 years ago, I saw it as speculative but possible. Now, in 2021, I see it as probable.

Whether it is or not will largely depend on how well we navigate the 2020s. Whether we truly take the path of utopia and abundance, or stumble down the path of destruction. We should all hold in our minds this evolution in consciousness as the carrot at the end of the decade.

Think about where we are in this early part of the 2020s. We are fully connected to a global fiber-optic and satellite, speed of light network – soon to be super-fast 5G – which enables us to stay connected and current with everything. We are now connecting without the limitations of time, distance and place. We are global and spatial in our human connectedness.

The degree of speed, connectivity, number of connected devices, explosion in the amount of data, and a geometric growth in the use of social media and apps are all accelerating. The speed of technological change today is the slowest it will be in your life-time.

We have the individuated sense of self in our physical reality, but our spatial reality is making us ever more one. One species globally. The acceleration of NOW brings us all ever closer together.

This causes cognitive dissonance at a level unprecedented in history. I am ME and simultaneously a part of a global WE. Something humanity has never experienced before.

What could possibly create cognitive dissonance more than an alteration of one's sense of self and how the new self-integrates into the emerging new reality? Moving from one location to another triggers a change in one's outlook and identity. Changing jobs or shifting into a new profession triggers a re-evaluation of who one is. Evolving from a solitary sense of self to a couple sense of self is disruptive. Vice versa – moving from an identity based upon union with another human to a solitary self is also disruptive.

In the 2020s we will not just be constantly adjusting, adapting and responding to rapidly changing externals, we will do the same with rapidly changing internals.

[1] The Shift Age book

[2] Entering the Shift Age book

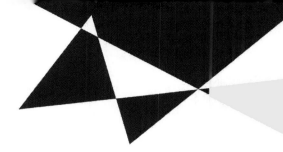

Chapter Nine – The Fork in the Road

The brilliant thinker, designer and futurist R. Buckminster Fuller ('Bucky') wrote several seminal books describing this pivotal moment in history, which he called the Fork in the Road. In "Utopia or Oblivion: Prospects for Humanity," published in 1969, Fuller wrote:

"Whether it is to be Utopia or Oblivion will be a touch-and-go relay race right up to the final moment… Humanity is in a final exam as to whether or not it qualifies for continuance in the Universe."

The 2020s is certainly the time for this final exam of our species. However, now that change is environmental, things are happening much faster than ever before. Relative to Fuller's fork in the road, we are there now. The actions we take now will set humanity's trajectory for the rest of this century.

In this decade we will have a clear choice: a conscious choice toward the future we want, or a mostly unconscious choice toward a disastrous future. By 2030 we will need to choose to:

- aggressively face our climate crisis particularly in eliminating all GHG emissions,

- decide to proactively guide how we want artificial/technological intelligence to develop,

- collectively deal with the science of human enhancement, DNA editing and human longevity,

- reinvent capitalism to focus it on the well-being of people and Nature,

- redesign and retrofit all the carbon-based infrastructure built throughout the 20th century,

- determine how best to deal with exponential technological growth,

- undertake informed population planning.

All of these issues are at a stage in the 2020s when we still have a choice, when we can create the future we want, when we can preemptively deal with big issues before they become wicked problems.

With our climate crisis, we have the opportunity to mitigate and prevent further tipping points that will accelerate planetary catastrophes beyond our capacity to adapt.

With the accelerating power of technological intelligence we can, at this time, decide which path is best for synergy with humanity rather than competition.

We can collectively decide how to deploy genetic engineering equitably for the benefit of all, rather than by enabling a dominant class of rulers.

We have increased global population by 300% in the last 70 years, mindlessly. It is in this decade we can proactively plan so that we don't over-populate to such a degree that society collapses.

It is in this decade that we can commit to planetary regeneration before it is too late.

The list could go on. The point is that we are at a fork in the road, a critical time in history when the future can still be chosen. To not proactively take advantage of this opportunity is to refuse to save ourselves from ourselves.

Cognitive Dissonance

We have discussed the degree of cognitive dissonance we can expect in the disruptive 2020s. Layer on top of all that an urgency to actively choose our collective future and this dissonance increases by an order of magnitude. In addition to just adapting – individually and collectively – we now have to purposely make significant decisions as to our future. The road not taken should be the mindless one we are currently on.

The 2020s will be the first decade when the limits of growth – in all areas – is an issue we can no longer put off. Never in the past have we had to make this type of decision on a global scale. There was always more room and more natural resources. Now our past has caught up with us. We have to shape our future now or it will shape us in negative ways.

The cognitive dissonance triggered by our rapidly changing reality is difficult enough to digest. We must at the same time chart a new course for our future. That is unprecedented.

We must personally learn to quickly adapt and increase our resilience. The same is true for society as a whole, governments and organizations. At the same time, we must consciously decide on the future we want and determine the actions we will need to take to get there. Humanity has never collectively done this, so there is no road map except for the sage advice of thinkers like Fuller, McLuhan, Toffler, and others.

We have to accept moral responsibilities for:

- our children, grandchildren and any subsequent generations,

- all animal species we can keep from extinction,

- all plant species we can keep from extinction,

- developing regenerative landscapes and technologies,

- ensuring that our ever deepening relationship with exponential technologies moves in the direction of an evolutionary step up,

- #humanityfirst and not the path toward losing our humanity,

- ensuring that opportunities for genetic enhancement and life extension move in a positive direction for all… not just the rich and powerful,

- redefining and reinventing capitalism to value people and planet above profit,

- retool democracy for the greatest good for the greatest number.

That is just my list. I am sure readers will easily think of many more additions to this.

This has to be a multi-generational effort. The silent and boomer generations have the time and resources to spend on this endeavor, but many suffer from 20th century legacy thinking. It is the Millennials, Digital Natives and Generation Alpha who will determine the success or failure of this most important effort. The youngest of these three generations will live past the year 2100, so they should have a say in what our collective reality will be for the next 75 years.

Just since I have started the final editing of this book, I have initiated an effort to proactively face our collective future. To discuss it here is risky since this initiative is just a few months old. This book will hopefully have years of viability in the marketplace. This new effort may not make it through the year 2021. Or it could take wing and have influence through the end of the decade. As a futurist, I feel impelled to take an action to help humanity navigate this fork in the road.

Many of us take the position that "someone else" or "some organization" is taking care of the wicked problems we face. Maybe we expect the government to use our taxes for enlightened, visionary efforts to determine the best way forward. There are organizations and activist groups committed to the common good, to a greener future, to a future of greater equality and abundance. Our situation is dire. Off-loading responsibilities to some group via tax deductible donations does not absolve anyone from taking further actions.

The very best quote on this is from Marshall McLuhan who, around the first Earth Day, said:

"There are no passengers on Spaceship Earth. We are all crew."

In fact, I took this quote as the genesis for co-writing the book, "This Spaceship Earth" and co-founding the nonprofit, This Spaceship Earth to create and spread "crew consciousness" to as many people as possible. En masse, we have created this planetary problem (albeit mostly unaware of the consequences of our actions). And, together, we can fix it… but we must be aware of what we want and conscious of our actions.

The point of facing the fork in the road this decade is that it is up to US.

I want to show how I am approaching this fork in the road reality for humanity in the 2020s. Perhaps what I am launching will resonate and you will join us. Perhaps you want to raise awareness and concern about this moment in time in another way. Either is fine, but taking action and, helping to raise awareness is something I implore all readers to do.

We all must understand that we are at the fork in the road, and we must act with urgency. Here is what I am doing:

As the year 2021 unfolded, I was in discussion with two global futurists whom I highly respect. We reviewed what we were seeing for the next ten years, and what we were planning on doing about it. We each referenced R. Buckminster Fuller. His fork in the road metaphor quickly came up and we all firmly agreed that humanity is there… now. After more discussion, we decided to launch the #forkintheroadproject. You can learn about it at https://forkintheroadproject.com/

Please join us or do something similar in your area of influence. It's an opportunity to design a future of well-being for us and our fellow species on Spaceship Earth.

In the next chapter we examine the global evolution in consciousness that must occur for humanity to survive and "graduate" to the next level.

Chapter Ten – Moving from Us/Them to WE

As noted in Book 1 of this series, humanity has entered into the global stage of human evolution. This is one of the most significant dynamics of the 2020s and will be the topic of a subsequent book in this series.

We have moved from family to tribe to village to city to city-state to nation state. Our only true remaining boundary is now global. Every aspect of human life is being integrated into an ever more connected global reality. The entire 170,000 year history of Modern Humanity has brought us to this next stage of our evolutionary journey. While our population exploded, man has migrated all over the planet. Add to this an ever-increasing electronic connectedness among people, and the result is an unprecedented degree of global integration.

We are now all, to a greater or lesser degree, global citizens. We travel everywhere. In 1950 the total number of global travelers, (defined as those traveling internationally, but not for business) was one million. In 2015 that number was one billion, a thousand-fold increase. Many people now travel thousands of miles in a day.

My son has spent about 70% of his post-college years traveling, studying and working outside the United States. He is an American who considers himself a global citizen. He is happily living and working in Amsterdam as a man with European sensibilities. He is far from alone. American Millennials have embraced the digital nomad lifestyle… something that will most certainly accelerate post-COVID-19. It is expected that by 2035 there will be one billion people living the nomadic work lifestyle, some 12% of humanity.[1]

The rapid acceleration to the Screen Reality triggered by COVID-19 means that working remotely is much more acceptable than it was at the end of the last decade. The 2020s decade will be much more about a distributed workplace that becomes global in scope. I can't imagine anyone reading this book who has not collaborated with, purchased from, or hired someone who lives in another country. We should remember how recent this phenomenon is. Traveling globally for business back in 2011, it was a major effort to find reliable high-speed wireless connections, and now, 4G and even early 5G is available almost anywhere.

Work is becoming ever more global.

Play is becoming ever more global.

Identity is becoming ever more global.

Consciousness is becoming ever more global.

This pushes us all into thinking of ourselves as global citizens. I have called myself that increasingly over the last 15 years as I delivered speeches in 16 different countries and on six continents. We are beginning to think of ourselves as WE globally. This is crucial for what we must face together in the 2020s. In all those presentations, and in the ones I currently deliver, I state that when I use the pronoun "we" I am referring to humanity. I do this not just for clarification, but because people typically assume I'm referring to a more limited group of people.

Increasingly people will move to the acceptance that "WE" is the human species. Extreme environmentalists take this even further and include all living species. The global stage of evolution has begun.

Our Big Issues and Challenges are Now All Global

Our climate crisis is global. We all live on Spaceship Earth. Sustainability is meaningless unless applied at the planetary level. It doesn't matter if a city, state or nation is sustainable or carbon neutral unless Spaceship Earth is.

There is no single nation state that can solve our climate crisis. At the very least, the top 20 GDP countries (who are also the top 20 GHG emitting countries) need to unite to successfully address CO2 emissions. These 20 countries – out of a total of 195 – consume 80% of the world's energy.[2] Humanity will fail our climate challenge unless these 20 countries can coordinate and decrease fossil fuel usage from 77% in 2019 to 30% in 2030.

Right now, the focus is on what nation states are doing. "Well, what about China?" "What about India?" are comments I often hear. Pointing the finger at another country isn't helpful. We are all in this together.

As Marshall McLuhan stated: *"There are no passengers on Spaceship Earth. We are all crew."* We can no longer live mindlessly as passengers. Or in a paradigm of Us vs. Them.

Our climate crisis requires that humanity thinks in terms of WE.

The Age of Intelligence is upon us. The rapidly accelerating development of Technological Intelligence is a global phenomenon. This is only now becoming clear. Historically, it was about which country had the fastest super-computer. More dangerously, countries are competing on deploying TI into weapons systems. This is to be expected when we think of ourselves as separate, competing nation states. Nation states think Us vs. Them.

Neuroscience, the final mapping of the brain that will take place in the 2020s, is about the science, not the country. For the species, not nation states. Neuroscientists collaborate globally, and would do more so if nation state issues of technological warfare were set aside.

Immigration and migration are global and planetary issues. There are few countries that are not dealing with this as a problem, one way or another. Back in 2012, I wrote about this[3.]

"The Shift Age will be the greatest age of migration in history. A greater number of humans will migrate at some time of their lives than in any prior age. Some of this will be due to the fact that the human population is larger than ever before. This means that the sheer number of us will be a factor. In addition, the percentage of the total population that will experience short-to-long term migration will increase as well."

We are becoming ever more assimilated globally. There are far more immigrants in North America and Europe as a percent of population today than when I was born.

In 2019 it was estimated that there will be between 50 to 100 million climate crisis refugees by 2030. That means that, over the next nine years, five million people per year will be relocating due to sea level rise, droughts, floods, or lack of arable land to grow food. The majority of these migrations will occur in the second half of the decade, so there is still some time to have a global discussion on how to handle this, but that must begin soon.

Migration and immigration are global issues that necessitate moving to a WE mindset.

Nation States

One of the biggest perpetuators of the Us vs. Them mindset is the nation state. Nationalism, along with the old, pre-enlightenment religions, are primary forces. For 15 years I have known that the 21st century would be looked back upon by future historians as the "end of the nation state century" (This assumes that in the 2020s we can rise up to successfully face our climate crisis. If not, there may not be historians at the end of this century].

In 2012 I wrote:

"...the nation state as we know it today was largely formed during the Industrial Age. During this age, the nation-state reached its zenith. The two-hundred-year period is winding down and giving way to the new global age.

This does not mean that the nation-state will go away. It is just that the concept and purpose of it are in a real process of change. As a number of countries experience the same issues and problems — financial interconnectedness, climate change, environmental degradation, vast amounts of capital flowing around the world at the speed of light, corporations that operate in dozens of countries, scarcity of adequate water and food... commonality becomes clearer through time. Global problems can only be solved by global solutions. No single nation-state or even a number of nation-states can solve global problems by themselves.

Of course the nation-state will remain in place. It still is the highest political human construct, at least for now. Nation-states will continue for decades. However, a transition is beginning to occur from a perception that the highest level of problem solving, the highest level of economic oversight, and the highest level of political philosophy is the nation-state, to a perception that perhaps the nation-state is too restricted and constricted to handle global problems.

Nation-states will continue, but will become more focused on the national issues of infrastructure, public safety, education, national economics, and the general well-being of the citizenry. In an ever more globally integrated culture, the nation-state will also take on the preservation of national culture, folk culture, and the cultural histories of the countries."

This trend will accelerate in the 2020s, and by 2040 will be well in place. Think of the magnitude of cognitive dissonance that will occur when humanity moves from thinking nationalistically to globally! Yes I am still an American, an Australian, a Russian or a South African… but I am increasingly a global citizen too. As a futurist I have been there for a decade, looking at the world as an American who is a global citizen. Again, the dual self becomes ingrained.

This movement is historically unprecedented as we have only entered the global stage of human evolution this century, in the Shift Age. The only part of history that rivals the disruption of this decade might be the 50 year transition from the Agricultural Age to the Industrial Age 1775-1825; and then again 100 years later after the Industrial Age North defeated the Agricultural South in the U.S. Civil War as mentioned in Chapter 5. The cognitive dissonance of moving from rural farming to urbanized production was significant, as was the 50 year period from 1875 to 1925 when the technologies of radio, telegraph, recording sound, internal combustion engine, airplanes and film became commonplace.

In the 2020s, at all levels of human activity, we will be undergoing unprecedented change. There will be few areas of human experience that will not result in cognitive dissonance.

Prepare yourself to live in this reality. There is no going back in the 2020s, only forward. How we individually and collectively process cognitive dissonance will be a key factor in whether we move in the direction needed to lay the foundation for the future we want to create… for us and for those who follow us in the decades ahead.

Our future is a global one. Each of us are both individuals and global citizens. This direction is inevitable in the long run. We must let go of the Us vs. Them mindsets that have plagued us throughout history. We have no model for a singular, global humanity. But WE must make one.

In some ways, this move from Us vs. Them will trigger the greatest amount of cognitive dissonance in the years ahead. We have all had times when we blamed "the other", when we couldn't marry because the intended was of another religion, when our country was being corrupted by "them". When we fought nationalistic wars we fought "them". When a family that was different from most of the neighborhood moved in, they were referred to as "them". When we are confronted by what our country is doing to face our climate crisis, we point to another country and say, "well, what about them?"

Every single war was about Us vs. Them. Moving to "WE" consciousness would go a long way toward ending war.

All of history up to the year 2000 was Us vs. Them. With the current Shift Age as the first step in the transition, the future history of humanity (hopefully by 2050), could be the WE era or WE period of civilization.

In the 2020s we must take the first steps toward that goal… with the knowledge that we will encounter significant cognitive dissonance, and the courage to forge ahead anyway.

The dawn of the WE era is now. Let's seize the day!

[1] https://observer.com/2020/11/life-in-2025-digital-nomads-will-change-travel-and-work-forever/

[2] Moving to a Finite Earth Economy

[3] Entering the Shift Age

Postscript

This is the second book in this series about the 2020s. As with the first book, this is a high-level high-concept book that can be read in two to three hours.

That is the idea, to write and publish a series of short books that provide high-level insights, forecasts and suggestions about living in this most disruptive of decades. Increasingly, we all live in a short attention span world. The most consistent comments about the first book, "The 2020s: The Most Disruptive Decade in History" was, "Not only did the book open my mind, it did so within three hours." And "I appreciate the quality combined with the brevity of the book. Exactly what should be published in this fast-paced world."

Here is the latest (tentative) list of titles, with the probable sequence:

"The 2020s: The Most Disruptive Decade in History" 2020

"The 2020s: A Decade of Cognitive Dissonance" 2021

"The 2020s: The Golden Age of Design and Redesign" 2021

"The 2020s: The Global Stage of Human Evolution" 2021

"The 2020s: Climate Crisis – Our Last Big Chance" 2022

"The 2020s: The End of Privacy" 2022

"The 2020s: The Beginning of 21st Century Finance" 2022

"The 2020: The Age of Intelligence"   2023

"The 2020s: A New Consciousness" 2023

The sequence and actual titles may change going forward. I ask only for a few hours of your time and an inquiring mind with each book.

Biography

David Houle is a futurist, speaker and strategist. Houle spent more than 20 years in media and entertainment. He worked at NBC, CBS and was part of the senior executive team that created and launched MTV, Nickelodeon, VH-1 and CNN Headline News.

Houle has won a number of awards. He won two Emmys, the prestigious George Foster Peabody Award and the Heartland Award for "Hank Aaron: Chasing the Dream" He was also nominated for an Academy Award. He is the Futurist in Residence at the Ringling College of Art + Design, the Co-Founder and Managing Director of The Sarasota Institute – A 21st Century Think Tank, the Honorary Futurist at the Future Business School of China and an initiator of the #forkintheroadproject.

He has delivered 1200+ speeches, presentations, and corporate retreats on six continents and sixteen countries. He is often called "the CEO's Futurist" having spoken to or advised 4,500+ CEOs and business owners in the past eleven years. He was invited to present "This Spaceship Earth" to scientists at the NASA Goddard Space Flight Center and at the EPA.

Houle coined the phrase the Shift Age and has written extensively about the future and the future of energy. This is his eleventh book. His primary web site is https://davidhoule.com/